Springer Proceedings in Mathematics & Statistics

Volume 169

More information about this series at http://www.springer.com/series/10533

Springer Proceedings in Mathematics & Statistics

This book series features volumes composed of select contributions from workshops and conferences in all areas of current research in mathematics and statistics, including OR and optimization. In addition to an overall evaluation of the interest, scientific quality, and timeliness of each proposal at the hands of the publisher, individual contributions are all refereed to the high quality standards of leading journals in the field. Thus, this series provides the research community with well-edited, authoritative reports on developments in the most exciting areas of mathematical and statistical research today.

Andrea Matta • Evren Sahin • Jingshan Li
Alain Guinet • Nico J. Vandaele
Editors

Health Care Systems Engineering for Scientists and Practitioners

HCSE, Lyon, France, May 2015

Springer

Editors
Andrea Matta
Shanghai Jiao Tong University
Shanghai, China

Jingshan Li
College of Engineering
University of Wisconsin
Madison, Wisconsin, USA

Nico J. Vandaele
Katholieke Universiteit Leuven
Leuven, Belgium

Evren Sahin
Laboratoire Génie Industriel
Ecole Centrale Paris
Châtenay-Malabry, France

Alain Guinet
DISP Laboratory
Lyon University, INSA de Lyon
Villeurbanne, France

ISSN 2194-1009 ISSN 2194-1017 (electronic)
Springer Proceedings in Mathematics & Statistics
ISBN 978-3-319-35130-8 ISBN 978-3-319-35132-2 (eBook)
DOI 10.1007/978-3-319-35132-2

Library of Congress Control Number: 2016951313

Printed on acid-free paper

This Springer imprint is published by Springer Nature
The registered company is Springer International Publishing AG Switzerland

Preface

This volume features selected and peer-reviewed contributions from the Second International Conference on Health Care Systems Engineering (HCSE 2015). This conference provides an opportunity to discuss operations management issues in health-care systems. The emphasis is on quantitative methods for the analysis, design and management of health-care systems.

The participants are faculties, students, medical doctors and chief nurses from several disciplines. The main objective is fostering the collaboration between operations management scientists and clinicians.

Scientists and practitioners have the opportunity to discuss about new ideas, methods and technologies for improving the operation of health-care organizations during a 3-day single session. The event emphasizes the research in the field of health-care systems engineering developed in close collaboration with clinicians.

This second conference took place in Lyon, France, between 27 and 29 May 2015 in the Saint-Joseph/St-Luc Hospital. A limited number of papers were selected under a double-blind review process. I would like to thank all of the Scientific Committee and the anonymous reviewers for the selection of the works. In total, 19 papers are included in the conference proceedings. Each paper was presented at the conference to all participants and discussed with experts from the clinical field.

I would like to express my deep gratitude to our invited speakers, Prof. Sally Brailsford for agreeing to address the conference on "Modeling Human Behavior in Healthcare Systems: Is It Possible and Why Should We Do It?" and Prof. Nico Vandaele with the topic of "The Vaccine Supply Chain Decathlon: The Reconciliation of Technology, Economy and Humanity". Their contributions are perfectly in line with the objective of the conference which tries to initiate collaborations between scientists and clinicians.

I would like to thank all the speakers, authors and discussants of the papers together with their accompanying persons for their participation in HCSE 2015.

I gratefully acknowledge the key people of the organization of this event: Hélène Grange, Michel Greco, Andrea Matta, Sylvie Meyran, Jérémie Leynon, Anastasie Schiffer, Samuel Vercraene…

My hope is that this conference will serve as a biennial forum for researchers, academics and clinicians in the broad area of health-care systems engineering to discuss their most recent research findings and to provide them with opportunities for project development and technology transfer.

Lyon, France Alain Guinet

Contents

Contributors

Daoud Ait-Kadi Department of Mechanical Engineering, Laval University, Ville de Québec, QC, Canada

Arianna Alfieri Politecnico di Torino, Torino, Italy

Carlo Banditori IBIS Lab, Dipartimento di Ingegneria Industriale, Università degli Studi di Firenze, Firenze, Italy

Seda Baş Industrial and Systems Engineering Department, Yeditepe University, Istanbul, Turkey

Emmanuel Beaudry Centre Hospitalier St-Joseph/St-Luc, Lyon, France

Paola Cappanera IBIS Lab, Dipartimento di Ingegneria dell'Informazione, Università degli Studi di Firenze, Firenze, Italy

Giuliana Carello Dipartimento di Elettronica, Informazione e Bioingegneria, Politecnico di Milano, Milan, Italy

Cecily Castelnovo Istituto di Matematica Applicata e Tecnologie Informatiche (IMATI), Consiglio Nazionale delle Ricerche (CNR), Milan, Italy

Wanying Chen DISP, INSA de Lyon, Bât. Jules Verne, Villeurbanne, France

Rui-Na Fan School of Economics and Management Sciences, Yanshan University, Qinhuangdao, Hebei, China

Estelle Forestier Centre Hospitalier St-Joseph/St-Luc, Lyon, France

Colin A. Graham Accident and Emergency Medicine Academic Unit, Chinese University of Hong Kong, Hong Kong, China

Hélène Grange Centre Hospitalier Saint-Joseph/Saint-Luc, Lyon Cedex, France

Alain Guinet DISP, INSA de Lyon, Bât. Jules Verne, Villeurbanne, France

Institut National des Sciences Appliquées de Lyon (INSA de Lyon), Villeurbanne, France

Johan Holmgren Faculty of Technology and Society, Department of Computer Science, Malmö University, Malmö, Sweden

Lori Hoschild University of Wisconsin Medical Foundation, Middleton, WI, USA

Alexander Hübner Department of Operations, Catholic University of Eichstätt-Ingolstadt, Ingolstadt, Germany

Afshin Jamshidi Department of Mechanical Engineering, Laval University, Ville de Québec, QC, Canada

M. Kakehi Department of Industrial System, Faculty of Symbiotic Systems Science, Fukushima University, Fukushima, Japan

M. Kawakami Faculty of Nursing, Jichi Medical University, Shimotsuke-shi, Tochigi, Japan

Sally Kraft Dartmouth-Hitchcock, Hanover, NH, USA

Heinrich Kuhn Catholic University of Eichstätt-Ingolstadt, Ingolstadt, Germany

Yong-Hong Kuo Stanley Ho Big Data Decision Analytics Research Centre, Chinese University of Hong Kong, Hong Kong, China

Paolo Landa Department of Economics and Business Studies, University of Genova, Genova, Italy

University of Exeter Medical School, Exeter (United Kingdom)

Ettore Lanzarone Istituto di Matematica Applicata e Tecnologie Informatiche (IMATI), Consiglio Nazionale delle Ricerche (CNR), Milan, Italy

Marie Lassaigne Centre Hospitalier St-Joseph/St-Luc, Lyon, France

Hyun-Rok Lee Industrial & Systems Engineering, KAIST, Daejeon, Republic of Korea

Hyo Kyung Lee University of Wisconsin-Madison, Madison, WI, USA

Taesik Lee Industrial & Systems Engineering, KAIST, Daejeon, Republic of Korea

Janny M.Y. Leung Department of Systems Engineering and Engineering Management, Chinese University of Hong, Hong Kong, China

Jérémie Leynon Centre Hospitalier Saint-Joseph/Saint-Luc, Lyon Cedex, France

Jingshan Li University of Wisconsin-Madison, Madison, WI, USA

Na Li Department of Industrial Engineering and Management, Shanghai Jiaotong University, Shanghai, China

Quan-Lin Li School of Economics and Management Sciences, Yanshan University, Qinhuangdao, Hebei, China

Elisabetta Listorti Politecnico di Torino, Torino, Italy

Azadeh Maroufkhani Université Grenoble Alpes, Grenoble, France

Maria Di Mascolo Université Grenoble Alpes, Grenoble, France

Andrea Matta Shanghai Jiao Tong University, Shanghai, China

Nadine Meskens UCL Mons, Louvain School of Management, Mons, Belgium

Sylvie Meyran Centre Hospitalier St-Joseph/St-Luc, Lyon, France

M. Miyake Course of Industrial Engineering and Management, Graduate School of Engineering, Kanagawa University, Kanagawa-ku, Yokohama, Japan

Zeynep Ocak Industrial and Systems Engineering Department, Yeditepe University, Istanbul, Turkey

Gaelle Olleon Centre Hospitalier St-Joseph/St-Luc, Lyon, France

Marie Persson Department of Computer Science and Engineering, Blekinge Institute of Technology, Karlskrona, Sweden

Samira Abbasgholizadeh Rahimi Department of Mechanical Engineering, Laval University, Ville de Québec, QC, Canada

Angel Ruiz CIRRELT, Université LAVAL, Québec, QC, Canada

Faculty of Business Administration, Université Laval, Pavillon Palasis-Prince 2325, rue de la Terrasse, Local 2423, Québec (Québec), G1V 0A6, Canada

Seren Schirra Centre Hospitalier St-Joseph/St-Luc, Lyon, France

Jeffery Sleeth University of Wisconsin Medical Foundation, Middleton, WI, USA

Michele Sonnessa Department of Economics and Business Studies, University of Genova, Genova, Italy

Inkyung Sung KAIST, Daejeon, Republic of Korea

M. Takanokura Department of Industrial Engineering and Management, Faculty of Engineering, Kanagawa University, Kanagawa-ku, Yokohama, Japan

S. Taki Department of Management Information Science, Faculty of Social Systems Science, Chiba Institute of Technology, Narashino-shi, Chiba, Japan

Elena Tànfani Department of Economics and Business Studies, University of Genova, Genova, Italy

Angela Testi Department of Economics and Business Studies, University of Genova, Genova, Italy

Filippo Visintin IBIS Lab, Dipartimento di Ingegneria Industriale, Università degli Studi di Firenze, Firenze, Italy

Manuel Walther Catholic University of Eichstätt-Ingolstadt, Ingolstadt, Germany

Tao Wang Institut National des Sciences Appliquées de Lyon (INSA de Lyon), Villeurbanne, France

Richard Welnick University of Wisconsin Medical Foundation, Middleton, WI, USA

Molly Williams University of Wisconsin Medical Foundation, Middleton, WI, USA

Semih Yalçindağ Industrial and Systems Engineering Department, Yeditepe University, Istanbul, Turkey

T. Yamada Department of Informatics, Graduate School of Information and Engineering, The University of Electro-Communications, Tokyo, Japan

Xiang Zhong University of Wisconsin-Madison, Madison, WI, USA

Systems Approach for Preventing Falls in Hospitals and Nursing Homes Using Sensing Devices Surrounding the Patient's Bed

M. Takanokura, M. Miyake, M. Kawakami, T. Yamada, S. Taki, and M. Kakehi

Introduction

Societies in Japan and other developed countries have been rapidly aging. Aging societies have also been recognized as a critical problem in developing countries. The population in Japan has decreased since 2012, and the elderly population, especially those 75 years in age and older, is critically increasing, whereas the working-age and younger population is continuously decreasing (Cabinet Office in Japanese Government 2015). A public nursing insurance system was adopted in Japan in 2000 to take measures against the nation's highly aging society. However,

M. Takanokura (✉)
Department of Industrial Engineering and Management, Faculty of Engineering, Kanagawa University, 3-27-1 Rokkakubashi, Kanagawa-ku, Yokohama 221-8686, Japan
e-mail: takanokura@kanagawa-u.ac.jp

M. Miyake
Course of Industrial Engineering and Management, Graduate School of Engineering, Kanagawa University, 3-27-1 Rokkakubashi, Kanagawa-ku, Yokohama 221-8686, Japan

M. Kawakami
Faculty of Nursing, Jichi Medical University, 3311-159 Yakushiji, Shimotsuke-shi, Tochigi 329-0498, Japan

T. Yamada
Department of Informatics, Graduate School of Informatics and Engineering, The University of Electro-Communications, 1-5-1 Chofugaoka, Chofu-shi, Tokyo 182-8585, Japan

S. Taki
Department of Management Information Science, Faculty of Social Systems Science, Chiba Institute of Technology, 2-17-1 Tsudanuma, Narashino-shi, Chiba 275-0016, Japan

M. Kakehi
Department of Industrial System, Faculty of Symbiotic Systems Science, Fukushima University, 1 Kanayagawa, Fukushima-shi, Fukushima 960-1296, Japan

A. Matta et al. (eds.), *Health Care Systems Engineering for Scientists and Practitioners*, Springer Proceedings in Mathematics & Statistics 169,
DOI 10.1007/978-3-319-35132-2_1

the social security costs in Japan are greatly increasing. To supplement the shortage of annual revenue against the expansion of social security costs, the government has decided to raise the consumption tax from 5 to 8 % in 2014, and to 10 % in 2017. This tax increase will be used for an enrichment of basic pensions, elderly medical care, long-term care, and childcare (Ministry of Health, Labour and Welfare Japan 2015). Elderly care is a social problem with higher priority in Japan.

An improvement in quality of life (QOL) has been one of the critical problems for elderly care. Nurses and care workers, referred to hereafter as care-givers, take care of patients and elderly persons, also referred to as care-receivers, for their daily activities. A devoted care-giver enhances the QOL and well-being of their care-receivers. Although the working-age population should be a resource for elderly care, such population is decreasing in Japan, as previously stated. There is an absolute lack in human resources for the elderly because the population distribution in Japan is highly distorted. In addition, a poor working environment for care-givers is regarded as a social problem (e.g., long working hours, low income, and high rate of care-givers leaving their jobs). Although these problems should first be solved politically, systems engineering can be a powerful tool for a socially based solution.

A fall is one of the most impactful types of accidents that can occur in a hospital or nursing home. It reduces the QOL of care-receivers and deteriorates the professional duties of care-givers. In addition, falls are regarded as a non-process type accident in hospitals and nursing homes. It is more difficult for care-givers to prevent a non-process type accident than a process type one because it cannot predict the occurrence of accident easily. Some serious accidents have been reported in Japan. For example, an elderly male fell around a bedside, and his neck was caught in a space between bed rails. Unfortunately, he was died after this accident. An elderly female stood up from a bed by grasping a bed rail, but it was not fixed sufficiently. She lost balance and fell. Her arm was fractured by this accident. Preventing falls should be accomplished through technical, organizational, and social measures. As a technical measure, some products for preventing falls are currently available in Japan. For example, touch sensors can be placed at a bedside or on the floor, and can notify care-givers if a care-receiver touches them when sitting on the bedside or standing up from the bed onto the floor. However, such sensing devices may issue a false alarm owing to their simple structure. In addition, video monitoring systems cannot be used for preventing falls in hospitals and nursing homes because of an inherent privacy issue. Recently, some micro-electro-mechanical systems (MEMSs) have been developed for monitoring daily activities and improving the QOL. For example, the activity meter with a tri-axial accelerometer has been used for the prevention of lifestyle-related diseases (Ohshima et al. 2012; Ohkawara et al. 2011; Tanaka and Tanaka 2009) and a quantitative understanding of child-care activities (Takanokura et al. 2014; Yamada et al. 2015). Nowadays, MEMS are regarded as an essential system for an enhancement of QOL in every stage of life.

In this study, we focus on falls around a bed, but accidents will occur in other environments such as walking. However, non-process type accidents are the critical problem in hospitals and nursing homes, and they occur at any time zone or at any place without care-givers especially at night. Patients and elderly persons first stand up from their bed before walking. Accidents in walking can be prevented by detection of dangerous motions while standing up.

We first applied MEMS technologies for preventing falls by detecting dangerous motions while standing up from a bed and when leaving a room in hospitals and nursing homes. In addition, a systems approach is necessary to manage such MEMS technologies during the nursing process. Thus, we discuss how MEMS technologies have been used effectively and efficiently for the prevention of falls in hospitals and nursing homes.

Objectives

First, we describe the development of our system for detecting and predicting care-receiver's motions using a micro-computer and sensing devices. This system detects the care-receiver's motions on and around their bed, and predicts dangerous motions that can lead to a fall. Second, we discuss how falls should be prevented by using measured data obtained from the developed system from the viewpoints of both care-givers and care-receivers. Finally, we model the nursing processes of care-givers in a hospital and nursing home, and consider how the developed system should be utilized into the nursing process for preventing falls.

Detection and Prediction of Dangerous Motions on and Around a Bed

Care-receivers have a higher occurrence of falls while standing up from a bed or when leaving their room. Before conducting such dangerous motions, the care-receiver raises their upper body from the bed, and then sits at the bedside. Such behavior may be a sign of a dangerous motion that could result in a fall. Before standing up from the bed, the pressure distribution around the care-receiver varies based on their motion. First, the care-receiver raises their upper body from the bed, thereby lowering the pressure around their head and shoulders. Then, the care-receiver moves to the bedside and sits. The pressure around the hips and legs is then reduced. We can detect this type of behavior from a continuous variation of pressure distribution on the bed by using a force sensing resistor (FSR) (No. 406, Interlink, USA) as a sensor to detect the care-receiver's motion. An FSR is rectangular in shape with 38.1-mm long sides. Furthermore, a fall will likely occur just as the care-receiver stands up from the bed. For example, a care-receiver with cognitive impairment may linger around their room, whereas those without such impairment may stand up from the bed in the dark to use the toilet during the night. For such care-receivers, we consider standing up from the bed and simply moving toward the door of the room as dangerous motions. Therefore, a passive infra-red human

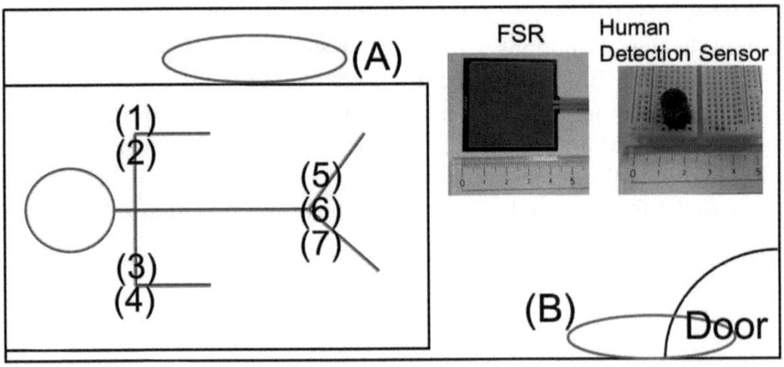

Fig. 1 Sensing devices and placement of them on and around a bed. FSRs are placed on *(1)–(7)*. Human detection sensors are set at *(A)* and *(B)*

detection sensor (NaPiOn spot type, Panasonic, Japan) is used for detecting these types of dangerous motions.

The placement of the sensing devices on and around a bed is shown in Fig. 1. When a care-receiver lies on their bed, their back and hips touch the mattress. There are different behaviors that occur when standing-up from a bed, but a popular one is the care-receiver first raising their upper body from the bed (sitting-up) and then sitting at the bedside. In this case, the pressure is first lowered around the shoulders and then around the hips. Therefore, four FSRs are placed under both sides of the shoulders, three FSRs are placed under the hips (chest), and two human-detection sensors are placed just under the bed (A) and in front of the door (B). The first four are activated when the care-receiver sits at the bedside, and the last one is activated when he or she opens the door to leave the room.

This system for preventing falls uses an *Arduino Uno*, which is a tool kit with an AVR microprocessor; seven FSRs; and two human-detection sensors. A secure digital (SD) memory card shield is also installed in the system. The data measured by the sensors are fed into the *Arduino Uno* and stored on the SD card.

The validity of the developed system was examined experimentally. It should be examined with participation of care-receivers in hospitals or nursing homes. However, the developed system is a prototype for prevention of falls, and safety of the system is not guaranteed sufficiently. Therefore, two healthy young males are selected as the participant. Participant A was 22 years old with a height of 170 cm and weight of 58 kg. Participant B was 22 years old with a height of 180 cm and weight of 66 kg. They first lied down on the bed, and then turned onto their left or right side or sat upright. After moving on the bed, they sat at the bedside, and then stood up from the bed. They then moved to the door. Figure 2 shows the measured signals of the FSRs and human-detection sensors for participant A. Period (1) was measured when the participant lied down on the bed and did not move. All data measured from the FSRs were equal to the maximal value (1023 arbitrary unit). The participant turned onto their right side during Period (3), and pressure at the left shoulder was reduced. Similarly, the pressure at their right shoulder was reduced

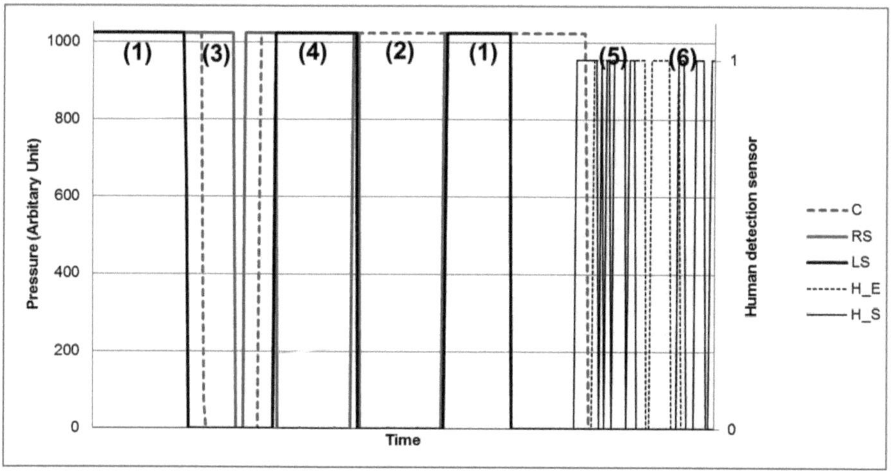

Fig. 2 Measured signals from the FSRs and human detection sensors for participant A. *C*, *RS*, and *LS* indicate the pressure data from the chest (hips), right shoulder, and left shoulder, respectively. The pressure is measured as an arbitrary unit. *H_E* and *H_S* indicate signals from human detection sensors placed in front of the door (entrance) and under the bedside, respectively. Sensors detect a value of "1" (existence) or "0" (absence)

during Period (4) because the participant turned onto their left side. The participant then sat up during Period (2), and thus the measured signals were reduced from both sides of the shoulder. During Period (5), the participant sat at the bedside and then stood up from the bed. None of the FSRs measured any pressure on bed, and the human-detection sensor under the bedside was activated. After standing up, the participant went to the door during Period (6), and the human-detection sensor in front of the door was activated.

As a result, we could detect and predict dangerous motions that could lead to a fall. In addition, we propose an algorithm for predicting falls through the use of the developed system, as shown in Fig. 3. The developed system can predict dangerous motions around a bedside and the door of a room, as shown in Fig. 3.

The system will notify the care-givers or care-receivers if alarm (A) or (B) is set off, as shown in Fig. 3, but the information provided depends on how the developed system is used. This issue is considered in the next section. In addition, dangerous motions are judged by the system when the threshold pressure value (600 or 200 in an arbitrary unit) is exceeded. This is a simple algorithm, and we intend to consider other algorithms such as a statistical measure, time-series analysis, or meta-heuristic approach (GA, neural network, etc.).

In this study, the participants are two healthy young males because of a safety reason. However, elderly persons in nursing homes have a different body shape such as high Body Mass Index compared with young males. In addition, patients and elderly persons move their body slowly. Therefore, we should consider proper threshold values or other algorithms for patients and elderly persons.

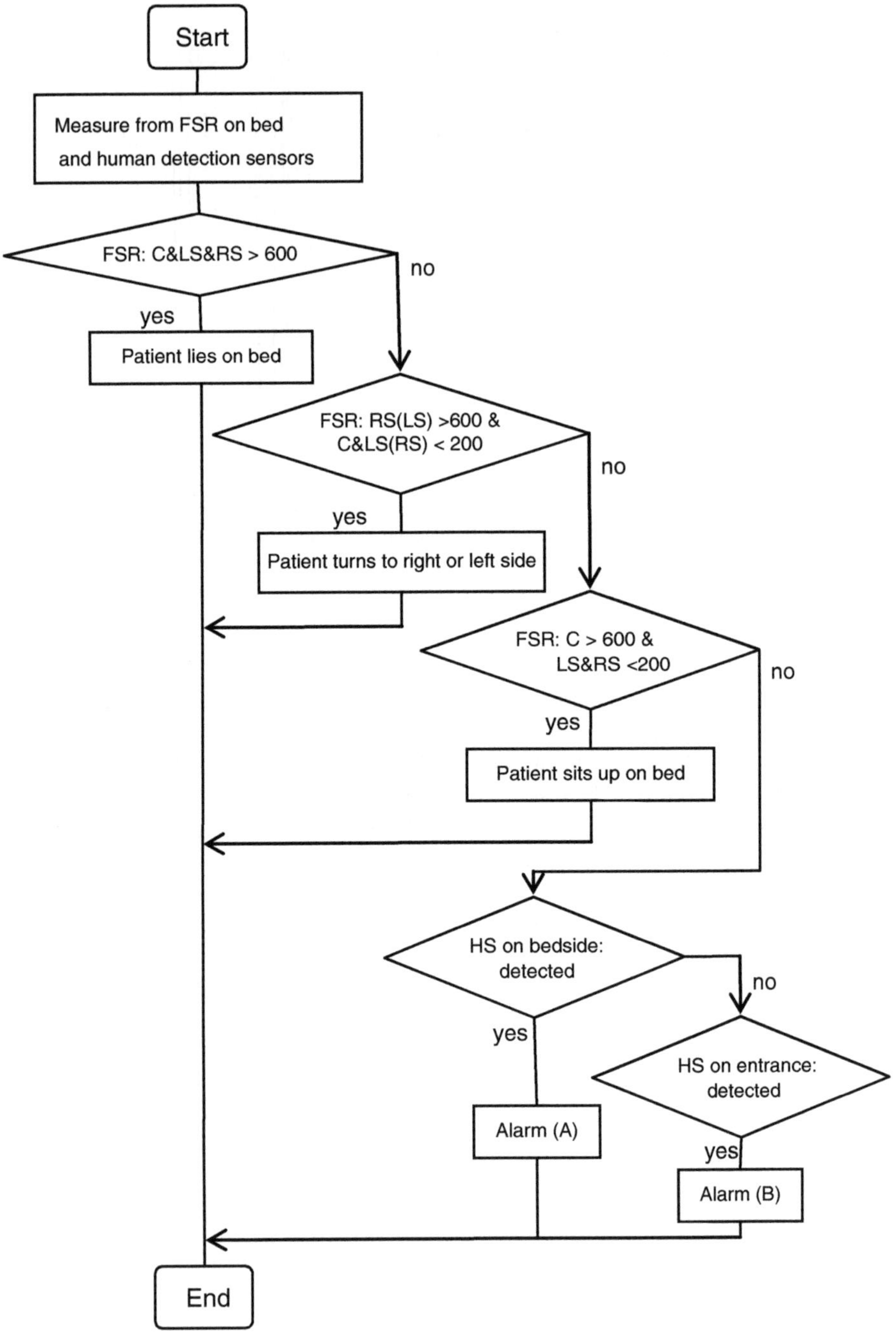

Fig. 3 Algorithm used for preventing a fall through the developed system. *C*, *LS*, and *RS* indicate the chest (hips), left shoulder, and right shoulder, respectively. *HS* indicates a human-detection sensor

Fall Prevention: Viewpoints of Care-Givers and Care-Receivers

The developed system can predict dangerous motions that may lead to a fall. The care-givers or care-receivers should be notified about a predicted motion by the system before a fall occurs. For care-givers, the notification should be through an alarm. For example, if a care-receiver with cognitive impairment intends to stand up near their bed and leave the room, the care-givers are notified of such motions by the system. After receiving an alarm from the system, the care-givers can reach the care-receiver as quickly as possible and help them restrain their dangerous motions. Such information should be provided to the care-givers through an emergency call, or should be recorded as a daily log during the nursing process. Recorded information is useful for care planning in hospitals and nursing homes. However, such information should be provided to not only care-givers but also to care-receivers. For example, after any surgical operation, a care-receiver should rest in bed and refrain from sitting up from the bed or standing. If they choose such a dangerous motion, the room light can be turned on or a warning message such as "*For your safety, do not stand up!*" can be given to the care-receiver. To implement this function into the developed system, we installed an LCD module (Fig. 4) and a loudspeaker into the system. The algorithm used for fall prevention (Fig. 3) provides an alarm (A) as a message to the care-receiver. At the same time, care-givers will receive a warning message from the system. Alarm (B) is not required for a care-receiver after an operation, but is needed for a care-receiver with a cognitive impairment. We can prevent falls from occurring by combining and judging the signals from the developed system (alarms (A) and (B) in Fig. 3) based on the symptoms or conditions of the care-receivers.

Fig. 4 LCD module installed in the developed system

Process Modeling of Care-Givers to Care-Receivers

Falls can be prevented not only through the use of technical products but also from organizational measures. As described previously, we developed the system to use sensing devices; however, this is a technical measure. It is also important for us to understand the nursing process of the care-givers when applying organizational measures to enhance the quality of healthcare. Therefore, we modeled the nursing processes of the care-givers to the care-receivers, and considered how the developed system should be utilized in these processes for preventing falls. A diagram for the process modeling was developed through the daily logs of care-givers taken at night. These daily logs were obtained from a hospital (cerebral surgery ward) and a long-term care facility.

Jun et al. proposed a node-link based characterization of eight different diagrams (Jun et al. 2009). A "flowchart" has the highest usability and utility among all of the diagrams used. The nursing process overall can be understood based on a flowchart. Although a care-giver's decisions were included in the flowchart, they could not be identified from daily logs. The "process content" and "state transition" diagrams have a higher usability and utility. The former can be used to understand a detailed task structure, and the latter can clarify a process in a care-receiver-centered manner. Therefore, we developed two diagrams from daily nursing logs taken at night in a hospital and a long-term care facility.

Figure 5 illustrates a process content diagram of nurses working in a hospital. The nurses have four processes: treatment, confirmation, preparation, and meetings. These processes are divided into respective sub-processes. The processes in the

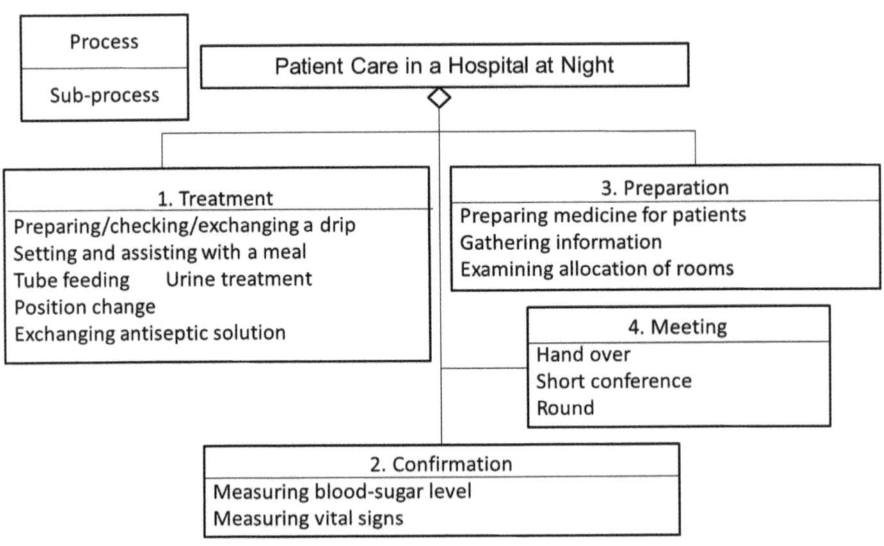

Fig. 5 Process content diagram of nurses in a hospital at night

diagram are the principal care (treatment, confirmation, and preparation) given to the patients and the duty (meetings) of the nurses. A long-term care facility simply takes care of elderly persons, and does not provide hospital treatment, and thus the "treatment" process is not included in the diagram. However, care workers are devoted to the care of elderly persons, and the nursing process was classified into confirmation, preparation, and meetings.

The process content diagram expresses the "healthcare process," which will enhance the QOL of care-receivers based on the duties of the care-givers. However, the QOL is reduced from unexpected accidents that occur in hospitals and nursing homes, which we call "non-process type accidents." Falls are typical cases of non-process type accidents. In addition, "process-type accidents" such as medication errors may occur in hospitals and nursing homes. A process content diagram is a powerful tool for the identification and prevention of process-type accidents, but non-process type accidents cannot be found from this type of diagram.

Figure 6 illustrates a state transition diagram of nurses in a hospital at night. The nursing process is expressed in a care-receiver-centered manner as a patient state. During the night, patients first receive medical care from nurses (drip, insulin injection, or tubal feeding) and an examination (measurement of vital signs). These processes are carried out during the early period of patient care at night, and non-

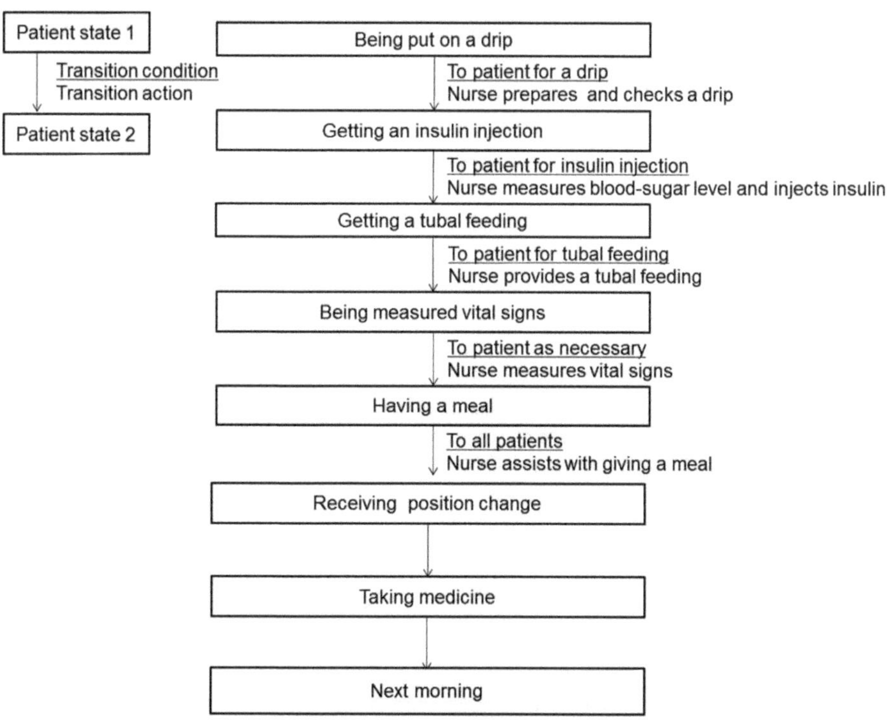

Fig. 6 State transition diagram in a hospital at night

process type accidents do not frequently occur because nurses can visually confirm the state of their patients. Nurses complete the process of patient care one time, and do not visually confirm the patient's state frequently after changing their position or giving them medicine. Non-process type accidents may therefore occur during this period of patient care. The developed sensing-device based system can be used for preventing falls after the early period of patient care at night in both hospitals and nursing homes. Although falls can be effectively and efficiently prevented using the developed system, we need to consider the implementation of the developed system as the next stage of this study.

Conclusion

We developed a system for preventing falls from occurring around a care-receiver's bed using MEMS technologies. Care-givers and care-receivers can be notified both visually and aurally if the system predicts dangerous motions on or around a bed. A state transition diagram indicates that falls may occur after an early care period at night because care-givers are unable to frequently confirm the care-receiver's state. Falls are a non-process type accident, and the developed system can be utilized during this period for the prevention of non-process type accidents. The system can improve the QOL of care-receivers and alleviate the professional duties of care-givers through a reduction in the number of falls. However, we should improve the ability of the system through the use of more complex algorithms, such as a meta-heuristic approach, and by analyzing the nursing process more precisely beyond the daily logs taken at hospitals and nursing homes.

Acknowledgements This research is partially supported by JSPS KAKENHI 26282091.

References

Cabinet Office in Japanese Government: Annual report on the aging society: 2013. http://www8.cao.go.jp/kourei/english/annualreport/2013/2013pdf_e.html. Accessed 30 Jan 2015

Jun, G.T., et al.: Health care process modelling: which method when? Int. J. Qual. Health Care **21**, 214–224 (2009)

Ministry of Health, Labour and Welfare Japan: The comprehensive reform of social security and tax—Why do we have to bear the burden? http://www.mhlw.go.jp/english/social_security/kaikaku_2.html. Accessed 30 Jan 2015

Ohkawara, K., Oshima, Y., Hikihara, Y., Ishikawa-Tanaka, K., Tabata, I., Tanaka, S.: Real-time estimation of daily physical activity intensity by a triaxial accelerometer and a gravity-removal classification algorithm. Br. J. Nutr. **105**, 1681–1691 (2011)

Ohshima, Y., Kawaguchi, K., Tanaka, S., Ohkawara, K., Hikihara, Y., Ishikawa-Takata, K., Tabata, I.: Classifying household and locomotive activities using a triaxial accelerometer. Gait Posture **31**, 370–374 (2012)

Takanokura, M., Ariizumi, K., Imai, N., Taki, S., Yamada, T.: Identification of high-intensity physical activities during child rearing using a triaxial accelerometer. In: Jang, R., Ahram, T. (eds.) Advances in Physical Ergonomics and Human Factors: Part II, pp. 93–102. AHFE Conference, USA (2014)

Tanaka, C., Tanaka, S.: Daily physical activity in Japanese preschool children evaluated by triaxial accelerometry: the relationship between period of engagement in moderate-to-vigorous physical activity and daily step counts. J. Physiol. Anthropol. **28**, 283–288 (2009)

Takanokura, M., Taki, S., Sato, T.: Measurement and analysis method for "Ikumen" activities including childcare tasks using a triaxial accelerometer J. Jpn. Ind. Manage. Assoc. **66**, 161–168 (2015)

A Multi-objective Patient Admission Planning Improving Resources Utilisation Under Bed Capacity Constraints

Alain Guinet, Nadine Meskens, and Tao Wang

Introduction

The hospital management takes place in an increasingly competitive environment.

Lots of hospitals have therefore decided to centre their organization on a best resources affectation and more precisely on one specific resource: the hospital beds.

Hospital bed management consists of managing as best as possible the welcoming capacity in order to take into account a volume of activity, while respecting quality criterion (quality of cares, delays, duration ...) under resources constraints.

Hospital bed management has an impact on the occupation rates of the different care units all along the patients' trajectory: from the beginning of the hospital stay, with an impact on the emergency service to the operating theatre, on intensive care units and even on rehabilitation centers and rest homes. Therefore, the hospital that wishes to optimize the allocation of its resources will have to cleverly schedule the admission of its patients. Indeed, during the patient admission, a bed in a ward is assigned and a series of resources (for medical diagnosis or a surgery) will be used.

In this paper, we try to assign an admission day and a bed to each elected patient who must be admitted to a surgery or a medical unit. A time window for hospitalization and a length of stay are associated to each patient admission. The unit capacity i.e. the number of beds of the medical and surgical units must be respected. Internal patient transfers between care units are allowed.

We take into account two objective functions. With the first one the smoothing of resources is sought. We have two kinds of resources. Diagnosis resources such

A. Guinet (✉) • T. Wang
Institut National des Sciences Appliquées de Lyon (INSA Lyon), DISP, Villeurbanne, France
e-mail: alain.guinet@insa-lyon.fr

N. Meskens
UCL Mons, Louvain School of Management, Mons, Belgium

A. Matta et al. (eds.), *Health Care Systems Engineering for Scientists and Practitioners*, Springer Proceedings in Mathematics & Statistics 169,
DOI 10.1007/978-3-319-35132-2_2

as X-rays and CT-scans could be used during the entire patient stay for surgical or medical unit admissions. Other resources are used only on the day of patient surgery such as operating rooms. We will call them "surgical resources." Whatever the resource, we will try to admit patients so that the use of resources is distributed as fairly as possible over days. Moreover the resource smoothing is also interesting to study the resource utilization rate (over or under utilization) and for resource sizing.

Regarding the second objective, we will try to allocate a bed in the unit of care where the surgical or medical operation will be performed on the patient. A patient who is not in the appropriate care unit linked to his pathology implies an organizational problem regarding the physicians' visits, a problem related to the nursing staff competencies, not always available to care for specific pathologies, and also problems regarding their internal transfer that can be costly, especially in terms of stretch-bearing activities, administration, etc. In addition, hospitals are in competition, satisfaction and patient safety become paramount and these are directly related to the quality of care. The cases where the patient is not in a suitable care unit happen when a care unit has no more beds available. We also found in real life some care units that were closed during weekends, also the patients were transferred to a ward where beds were vacant. Internal patient transfers are thus realized.

A cost penalty per period (day) will be used when the patient will be in an inadequate care unit. We will therefore seek to ensure that the patient will spend the least amount of time as possible in such a care unit. Furthermore, it must be ensured that hospitalization occurs during the window of admission. Indeed the physician will give a date not earlier than and a date not later than for the admission. When the hospitalization window is not respected, penalty costs will be taken into account.

This paper is structured into four parts: after the introduction the second part of this paper presents a review of literature linked to the studied problems and helps us in developing the model which was built for decision makers. A third section related to hospital bed management model follows it. The fourth part of this dissertation presents an experiment of our model based on field data.

Previous Works

Decision making regarding bed management, could be classified according to the scale and the details of decision problems modelling: bed sizing, bed organization, bed occupancy and admission or bed planning.

A lot of researchers were interested in determining the optimal number of beds in hospitals but also in recovery rooms, intensive care units or emergency departments. Let us quote some of them as examples: Li et al. (2009), Ma and Demeulemeester (2012), De Bruin et al. (2009), Cochran and Roche (2009), Kokangul (2008), and Troy and Rosenberg (2009).

Less numerous are studies regarding bed organization such as the assessment of different strategies for bed reservation (Kim and Horowitz 2002) or to allocate beds among hospital wards in order to minimize the over-booking (Holm et al. 2013)

Bed occupancy is important to be studied but is also impacted by a relevant management of care units or other departments such as the operating theatre. Indeed, building cycle master surgery schedules influences the bed occupancy rates. Some papers investigated models related to bed occupancy and consider operating theatre occupancy as a resource that needs to be taken into account (Guinet and Chaabane 2003; Wargon et al. 2014).

When a patient is admitted, a bed in a care unit is granted for them. The problem called bed planning is also called admission planning or admission scheduling. Bed planning consists mainly of assigning a bed to each elective patient and each post-acute patient i.e. a patient who leaves the emergency department in order to be hospitalized. Some authors were interested in the problem of bed planning: Hulshof et al. (2013), Ben Bachouch et al. (2012), and Gartner and Kolisch (2014).

In conclusion we have observed that most of previous works focus on bed sizing. Bed occupancy has been mainly studied by operating theatre management investigations. It could be interesting to do the reverse i.e. to study surgical and medical resources by a bed management approach. That is the aim of this paper.

To the best of our knowledge no research has sought to smooth the resources used throughout the patient's stay or has made a distinction between the surgical resources and the diagnosis resources. Furthermore we seek to optimize two objective functions. The first is to maximize satisfaction and patient safety by giving them a bed in the appropriate care unit and during the required time window. Internal patient transfers have not been taken into account in bed planning literature. The second objective seeks to smooth the resources. We have not found other research that realized this.

In conclusion according to the literature shortage on admission or bed planning worthwhile investigations could be done in this area.

Model

At the hospital scale, the different care units are schematized with a bed capacity. Resources like X-rays and CT scans needed by the patients are defined by their available times. Considering that a patient stay may be split on more than one care unit, the aim of the model is to balance the workload of resources (diagnosis resources and surgical resources) authorizing the location of the patient in another service than the intended one. Indeed, we search to reduce the possible bottlenecks of conflict resources used at the same time by several patients.

We will differentiate patients who enter the hospital for a surgery (called surgical patients) and those who go to receive care in medicine (called medical patients). The surgical patients use diagnosis resources as well as surgical resources, whereas the medical patients need only diagnosis resources.

To avoid too many wrong patient locations, another objective tries to minimize the number of days in an inadequate care unit. This objective function will also penalize heavily (leading to non-feasibility) patient stays that would not be planned in the hospitalization time window. The penalty value is fixed at 1 for each day spent in an inadequate care unit and at a high value if the hospitalization window is not respected.

Notation

I	Set of patients, $i = 1, 2, \ldots N = IS \cup IM$
IS	Set of surgical patients, $s = 1, 2, \ldots S$
IM	Set of medical patients, $m = 1, 2, \ldots M$
J	Set of care units, $j = 1, 2, \ldots U$
Γ	Set of diagnosis resources, $k = 1, 2, \ldots K$
R	Set of surgical resources, $q = 1, 2, \ldots Q$
P	Set of periods (days), $t = 1, 2, \ldots T$
a_i	Early hospitalization date for patient i
b_i	Latest hospitalization date for patient i
d_i	Length of stay for patient i (in days)
cu_i	Care unit required by patient i
m_{jt}	Number of beds available in care unit j at period t
rs_{sq}	Number of minutes required by a surgical patient for surgical resource q
rd_{ik}	Number of minutes required by patient i for diagnosis resource k
c_{ijt}	Penalty cost of hospitalization stay in care unit j for patient i during period t:

$$\begin{cases} +9999 & \text{If } t < a_i \text{ or } t > b_i + d_i - 1 \text{ (if admission day is outside the patient admission window)} \\ 1 & \text{If } j \neq cu_i \text{ and } a_i \leq t \leq b_i + d_i - 1 \text{ (if patient is admitted in the non} - \text{required care unit)} \\ 0 & \text{otherwise} \end{cases}$$

Decision Variables

z_{ijt}	Binary variable equal to 1 if patient i is hospitalized in care unit j during period t, 0 otherwise
x_{it}	Binary variable equals to 1 if patient i starts his stay at t, 0 otherwise
$Cmax_k$	Maximum number of minutes required for diagnosis resource k
$Cmax_q$	Maximum number of minutes required for surgical resource q

Model

$$Min\,(Z_1) = \left[\left(\sum_{k=1}^{K} \mathrm{Cmax}_k + \sum_{q=1}^{Q} \mathrm{Cmax}_q\right)\right] \tag{1}$$

$$Min\,(Z_2) = \left[\left(\sum\nolimits_{i=1}^{N} \sum\nolimits_{j=1}^{U} \sum_{t=1}^{T} z_{ijt} * c_{ijt}\right)\right] \tag{2}$$

Subject to

$$\left[\left(\sum_{j=1}^{U} \sum_{i=1}^{N} z_{ijt} * rd_{ik}/d_i \leq \mathrm{Cmax}_k\right)\right] \quad \forall t \in \{1, \ldots, T\} \quad \forall k \in \{1, \ldots, K\} \tag{3}$$

$$\left[\left(\sum_{\substack{s=1 \\ s \in IS}}^{S} \left(x_{st-1} * rs_{sq}\right) \leq \mathrm{Cmax}_q\right)\right] \quad \forall t \in \{2, \ldots, T\}, \forall q \in \{1, \ldots, Q\} \tag{4}$$

$$\left[\left(\sum_{i=1}^{N} z_{ijt} \leq m_{jt}\right)\right] \quad \forall t \in \{1, \ldots, T\}, \quad \forall j \in \{1, \ldots, U\} \tag{5}$$

$$\left[\left(\sum_{j=1}^{U} z_{ijt} \leq 1\right)\right] \quad \forall i \in \{1, \ldots, N\}, \quad \forall t \in \{1, \ldots, T\} \tag{6}$$

$$\left[x_{it} \geq \sum_{j=1}^{U} z_{ijt} - \sum_{j=1}^{U} z_{ijt-1}\right] \quad \forall i \in \{1, \ldots, N\}, \quad \forall t \in \{1, \ldots, T\} \tag{7a}$$

$$\left[\sum_{t=1}^{T} x_{it} = 1\right] \quad \forall i \in \{1, \ldots, N\} \tag{7b}$$

$$z_{ij0} = 0, \quad \forall i \in \{1, \ldots, N\}, \quad \forall j \in \{1, \ldots, U\} \tag{7c}$$

$$\left[\left(\sum\nolimits_{j=1}^{U}\sum_{t=1}^{T} z_{ijt} = d_i\right)\right] \quad \forall i \in \{1, \ldots, N\} \tag{8}$$

$$z_{ijt}, x_{it} \in \{0, 1\} \quad \forall i \in \mathrm{I},\ \ \forall j \in \{1, \ldots, U\},\ \ \forall t \in \{1, \ldots, T\} \tag{9}$$

Regarding the model, on one hand a smoothing of resource demands is required and on the other hand fitting patient assignments to adequate care units is sought. So, we are faced with two objectives functions. The first one (1) smoothes resource demands and is expressed in minutes and the second one (2) minimizes the penalty cost. This latter integrates a high cost if the hospitalization of patients is outside the admitted period of hospitalization and a low cost due to the inappropriate patient assignments to care units. Putting a high value (9999) to a placement outside the window, will force the linear program to respect the earliest and the latest dates as hard constraints. We should not get this value for Z_2 unless there is no other feasible solution. For the other part of penalty costs, it represents the number of days that patients are not in the right care unit.

The constraints (3) and (4) allow us to calculate the maximum demands of resource for each period respectively for diagnosis resources and surgical resources. The constraints (3) concern all patients and focus on diagnosis resources. The latter can be used at any time during the patient's stay. On the other hand the constraints (4) focus on surgical resources and therefore relate only to surgical patients. In addition, surgical resources are used only on the day of surgery (that is to say, the day following the patient's admission). The constraints (5) ensure that the bed capacity in care units at period t is respected. The constraints (6) specify that each patient be hospitalized at most in one care unit per day. The constraints (7a)–(7c) prohibit non continuous patient stay as illustrated hereafter. From period 4 to 8, a patient stay of 3 days modelled by a stream of five z binary variables 01011 is forbidden because two x binary variables must be equal to one (due to (7a)) that is proscribed by (7b). A patient stay modelled by 01110 is allowed because only one x binary variable is set to 1. The constraints (8) enable both to assign a bed and to respect the length of stay for each patient. Decision variables z_{ijt} and x_{it} are binary variables (9).

Multi-objective Procedure

We implemented a multi-objective procedure to gradually introduce the objectives into our model. This procedure consists of solving a first objective. Once the optimal value of this objective is identified, this objective is inserted into the model as an additional constraint. The second objective is then optimized. For example, to minimize Z_1 under a set of constraints C while considering that the optimal value of

Z_1 is OPT, the constraint $Z_1 \leq OPT$ is added to the set of constraints C and the next objective is used as the function that needs to be optimized.

Since we have two objective functions without particular priority for one or the other, we will proceed in two steps: first optimize Z_1 and next Z_2, and inversely.

Experiment

Solving three problems sizes assesses the model quality: 40 patients (PT), 80 patients (MT) and 120 patients (GT). For each problem size 10 tests are generated. Regarding to the two multi-objective procedures, 60 problems are solved.

Data

The study horizon is equal to 28 days. One surgical resource and two diagnosis resources are considered: operating rooms, scans and X-rays. Three medical care units are retained: internal medicine, surgery and orthopedic surgery. The last two hospital specialties require operating rooms. A uniform law is employed to define the care unit required by a patient. Table 1 specifies the statistical laws employed to generate surgical and diagnosis resource demands. Strum et al. (2000) have shown that log-normal law outperforms normal law to generate surgery processing times. In our experiment, PT set represents patients with great length of stay and GT set regroups patients with short length stays. The distribution of length of stay (LOS) is asymmetrical and can be fitted by the Lognormal, the Weibull or the Gamma law distributions (Marazzi et al. 1998) Weibull law or Gamma law give similar results. If the short-tailed model is considered, Weibull law must be used, otherwise log-normal law. Log-normal law has been retained for our experiment.

For warm-up reasons, the number of available beds per care unit is fixed to 4 for the first period, to 6 for the second period and to 8 for the next periods. Early hospitalization dates belong to interval $[1, 28 - d_i + 1]$ and are generated with a uniform law. Latest hospitalization dates b_i are lower than 28 and greater or equal to $a_i + \alpha$. α is generated with a uniform law between 0 and 7. Table 2 specifies also the law parameters for generating the lengths of patient stay.

Results

The mathematical model was computed with CPlex/oplrun using the OPL modeling language. A computer with an Intel Core i7 2.7 GHz CPU and 8 GB of allocated RAM gives the following results. An R script randomly generates all entry data of the OPL model.

Table 1 Statistical laws and their parameters for diagnosis resources

	Operating room	CT scan	X-rays
Distribution law	Log normal LN2(μ,σ)	Log normal LN2(μ,σ)	Log normal LN2(μ,σ)
Ranges		[1/6,3] h	[1/6, 3] h
Parameters for PT problems	$\mu = 60$ min	$\mu = 80$ min	$\mu = 40$ min
	$\sigma = 20$	$\sigma = 20$	$\sigma = 15$
	Range: [1, 2] h	(2 scans)	(2 X-rays)
Parameters for MT problems	$\mu = 120$ min	$\mu = 80$ min	$\mu = 40$ min
	$\sigma = 40$	$\sigma = 20$	$\sigma = 15$
	Range: [2,4] h	(2 scans)	(2 X-rays)
Parameters for GT problems	$\mu = 60$ min	$\mu = 40$ min	$\mu = 30$ min
	$\sigma = 10$ min	$\sigma = 1/6 \times 40$	$\sigma = 5$ min
	Range: [1/2, 2] h	(1 scan)	(1 X-ray)

Table 2 Statistical laws and their parameters for patient stays

	Early hospitalization date a_i	Length of stay d_i
Distribution law	Uniform law	Log normal LN2(μ,σ)
Parameters for all problems	$[1, 28 - d_i + 1]$ days	$\mu = 5$ days
		$\sigma = 1/6 \times 5$

Using the same 30 datasets, the tests are organized in a way that the objective functions Z_1 and Z_2 are optimized respectively as it should be in any classical optimization problem solving, followed by a multi-objective procedure as explained in the section "Multi-objective Procedure".

Smoothing the Resource Demands: Minimizing Z_1

Firstly, the objective function Z_1 is activated to smooth the amount of demands for different diagnosis resources and surgical resources. For each of the three problem sizes, 10 different datasets are generated according to the analysis in section "Data". A total of 30 problems are solved.

In the following experimental results tables, the optimal values of the active objective function are in bold type. Parallel values of another inactive objective function are also extracted in order to make a comparison between tests.

As shown in Table 3, if hospital decision makers (especially those who take charge of diagnosis or surgical resources, and ignore patient assignment in inpatient care units) focus only on the optimization of resource utilization in their own departments, the penalty costs from inappropriate assignments will increase.

Table 3 Experimental results by minimizing Z_1

	40 Patients (PT)		80 Patients (MT)		120 Patients (GT)	
Test	Z_1	Z_2	Z_1	Z_2	Z_1	Z_2
1	**566**	311	**1158**	316	**805**	254
2	**526**	275	**1083**	258	**861**	232
3	**675**	299	**1149**	285	**859**	215
4	**702**	255	**1127**	270	**836**	215
5	**439**	329	**1226**	260	**882**	259
6	**516**	310	**1241**	253	**841**	241
7	**476**	298	**1165**	302	**890**	242
8	**512**	288	**1187**	309	**838**	261
9	**570**	286	**1054**	254	**854**	224
10	**521**	312	**1173**	267	**892**	229

Table 4 Experimental results by minimizing Z_2

	40 Patients (PT)		80 Patients (MT)		120 Patients (GT)	
Test	Z_1	Z_2	Z_1	Z_2	Z_1	Z_2
1	1240	**1**	2119	**18**	1360	**0**
2	1089	**0**	1736	**33**	1257	**0**
3	1190	**15**	1756	**9**	1250	**0**
4	1003	**0**	1724	**15**	1393	**0**
5	807	**0**	1856	**1**	1301	**5**
6	1155	**1**	2006	**2**	1152	**0**
7	721	**1**	2437	**0**	1348	**0**
8	980	**0**	1689	**2**	1179	**0**
9	1059	**0**	1685	**8**	1438	**0**
10	722	**0**	1779	**29**	1346	**0**

Minimizing the Penalty Cost Z_2

Secondly, the objective function Z_2 is activated to make an appropriate patient assignment to bed with a minimum penalty cost. Using the same datasets, a total of 30 problems are solved.

The experimental results in Table 4 present an inverse scenario, decision makers in inpatient care units could make a perfect patient assignment that avoids all unnecessary penalty costs, but at the expense of overcharging medical diagnosis services and operating rooms.

In order to have an approximated value of Z_1 and Z_2, we can estimate the upper bound of these two objective functions.

The maximum value of Z_2 (if the time windows are respected) for 40 patients is 1120 ($= 40 \times 28$); for 80 patients, 2240 and for 120 patients, 3360. It represents the maximum number of days when all patients would be placed during their entire stay in an inappropriate care unit.

In Table 3, we see that the proportion of days spent in an inappropriate care unit decreases dramatically when the number of patients increases. For 40 patients this represents an average of 26 %, for 80 patients 12 % and for 120 patients 7 %. When Z_2 is first minimized (Table 4), these values are nearly zero.

As each patient uses one CT scan and one X-rays and assuming that about a half of patients will go to the operating room, an approximation of a higher bound of Z_1 for 40 patients is equal to $40 \times (60+20) \times 1/2 + 40 \times (80+20) + 40 \times (40+15) =$ 7799 min. This approximation represents also a maximum of the resource utilization time. For 80 patients, this approximation is 18,800 min and 14,000 min for 120 patients.

Regarding to 40 (80; 120) patients, seeking to smooth the maximum use of resources (Table 3), we obtain an average rate of use around to 7 % (6, 15 %; 6, 1 %) of the maximum value. When we first minimize Z_2 (Table 4), this gap increases to 9 % (10 %; 9 %).

Multi-objective Optimization

Global optimization of hospital resources requires meeting local optimization needs of all concerned care units. Moreover in some cases, a compromise should be found. According to our multi-objective procedure, the optimal values of Z_1 in Table 3 are transferred into a new set of constraints $Z_1 \leq$ OPT, Z_2 is then optimized.

Table 5 gives the results when Z_2 is optimized after Z_1. The penalty costs decrease considerably compared with the results in Table 3, even though they are a bit higher than the optimal values in Table 4 for the last three tests of 40-patients problem, five tests of 80-patients problem and the last test of 120-patients problem.

To a comprehensive analysis of these two objective functions, the order that they are optimized is inversed. The optimal values of Z_2 in Table 4 are transferred into a new set of constraints $Z_2 \leq$ OPT, Z_1 is then optimized.

In Table 6, the amount of resource demands is reduced by more than one-third of the values shown in Table 4.

Table 5 Experimental results by minimizing Z_2 after Z_1 optimized

	40 Patients (PT)		80 Patients (MT)		120 Patients (GT)	
Test	$Z_1 \leq$	$\mathbf{Z_2}$	$Z_1 \leq$	$\mathbf{Z_2}$	$Z_1 \leq$	$\mathbf{Z_2}$
1	566	**1**	1158	**18**	805	**0**
2	526	**0**	1083	**35**	861	**0**
3	675	**15**	1149	**10**	859	**0**
4	702	**0**	1127	**16**	836	**0**
5	439	**0**	1226	**2**	882	**5**
6	516	**1**	1241	**2**	841	**1**
7	476	**1**	1165	**1**	890	**0**
8	512	**2**	1187	**2**	838	**0**
9	570	**2**	1054	**8**	854	**0**
10	521	**2**	1173	**29**	892	**1**

Table 6 Experimental results by minimizing Z_1 after Z_2optimized

	40 Patients (PT)		80 Patients (MT)		120 Patients (GT)	
Test	$\mathbf{Z_1}$	$Z_2 \leq$	$\mathbf{Z_1}$	$Z_2 \leq$	$\mathbf{Z_1}$	$Z_2 \leq$
1	**630**	1	**1326**	18	**835**	0
2	**557**	0	**1107**	33	**849**	0
3	**674**	15	**1207**	9	**862**	0
4	**668**	0	**1185**	15	**846**	0
5	**431**	0	**1313**	1	**882**	5
6	**504**	1	**1308**	2	**852**	0
7	**468**	1	**1187**	0	**905**	0
8	**530**	0	**1282**	2	**852**	0
9	**597**	0	**1227**	8	**870**	0
10	**546**	0	**1157**	29	**903**	0

If we compare the results in Tables 5 and 6, we see that a small improvement on Z_2 leads to a deterioration on Z_1.

Another finding is that when there is no change in Z_2, Z_1 increases in value. Considering from the viewpoint of optimizing hospital's resource utilization in its entirety, it is more efficient to minimize Z_1 before Z_2 than the reverse.

Conclusion

We develop a model to schedule the admission of patients trying to smooth the utilization of diagnosis resources such as CT-scans and of surgical resources such as operating rooms. We also allocate a bed to each patient if possible in an adequate care unit taking into account the bed capacity of each care unit. We face a multi-objective problem with two conflicting criteria: to minimize the penalty cost and to smooth the resources demands. A lexicographic optimization was considered to solve this problem. It means that an order of importance was set between the objectives: the first objective is optimized; then the second one is optimized under the constraint that the first one stays optimal. In the future we want to optimize simultaneously both objectives by way of finding out a set of Pareto optimal solutions. The manager can thus choose among them the solution he/she prefers and estimates the trade-offs between the possible solutions.

References

Ben Bachouch, R., Guinet, A., Hajri-Gabouj, S.: A mixed linear model for hospital bed planning. Int. J. Prod. Econ. **140**, 833–843 (2012)

Cochran, J.K., Roche, K.T.: A multi-class queuing network analysis methodology for improving hospital emergency department performance. Comput. Oper. Res. **36**(5), 1497–1512 (2009)

De Bruin, A.M., Bekker, R., Van Zanten, L., Koole, G.M.: Dimensioning hospitals wards using the Erlang loss model. Ann. Oper. Res. **178**(1), 23–43 (2009)

Gartner, D., Kolisch, R.: Scheduling the hospital-wide flow of elective patients. Eur. J. Oper. Res. **233**, 689–699 (2014)

Guinet, A., Chaabane, S.: Operating theatre planning. Int. J. Prod. Econ. **85**, 69–81 (2003)

Holm, L.B., Luras, H., Dahl, F.A.: Improving hospital bed utilization through simulation and optimization with application to a 40% increase in patient volume in a Norwegian general hospital. Int. J. Med. Inform. **8**, 80–89 (2013)

Hulshof, P.J.H., Boucherie, R.J., Hans, E.W., Hurink, J.L.: Tactical resource allocation and elective patient admission planning in care processes. Health Care Manag. Sci. **16**, 152–166 (2013)

Kim, S.-C., Horowitz, I.: Scheduling hospital services: The efficacy of elective surgery quotas. OMEGA Int. J. Manag. Sci. **30**, 335–346 (2002)

Kokangul, A.: A combination of deterministic and stochastic approaches to optimize bed capacity in a hospital unit. Comput. Methods Programs Biomed. **90**(1), 56–65 (2008)

Li, L.X., Beullens, P., Jones, D., Tamiz, M.: An integrated queuing and multi-objective bed allocation model with application to a hospital in China. J. Oper. Res. Soc. **60**, 330–338 (2009)

Ma, G., Demeulemeester, E.: A multilevel integrative approach to hospital case mix and capacity planning. Comput. Oper. Res. **40**(9), 2198–2207 (2012)

Marazzi, A., Paccaud, F., Ruffieux, C., Beguin, C.: Fitting the distributions of length of stay by parametric models. Med. Care **36**, 915–927 (1998)

Strum, D.P., May, J.H., Vargas, L.G.: Modeling the uncertainty of surgical procedure times: comparison of log-normal and normal models. Anesthesiology **92**, 1160–1167 (2000)

Troy, P.M., Rosenberg, L.: Using simulation to determine the need for ICU beds for surgery patients. Surgery **146**, 608–620 (2009)

Wargon, M., Taright, N., Casalino, E., Pateron, D., Guidet, B.: A simulated discrete-event and queuing model to reduce transfers from the Emergency Department and to Optimize Hospital Bed Management. Adv. Emerg. Med. 7 pages (2014). http://dx.doi.org/10.1155/2014/478675

Multi-criteria Decision Making Approaches to Prioritize Surgical Patients

Samira Abbasgholizadeh Rahimi, Afshin Jamshidi, Angel Ruiz, and Daoud Ait-Kadi

Introduction

After patients are referred to surgery, their situation is examined. If patients have non-life threatening condition, they will be enrolled on a first-come, first serve basis. But if their conditions are life-threatening they will be registered on a priority waiting list. Higher priority-patients will be selected for service prior to those with a lower priority, regardless of when they are placed on the list (Randolph 2013). Prioritization of patients on waiting lists and their access to treatment based on various factors is one of the major issues within healthcare organizations. According to US National Library of Medicine (2012) this access is defined as "the degree to which individuals are inhibited or facilitated in their ability to gain entry to, and to receive care and services from the health care system". In many medical procedures, waiting times effect directly on patients' health and quality of care. Reports identify harms related to long wait times, including poorer medical outcomes from care and an increased risk of adverse events. For some diseases, delayed treatment can cause reduction in effectiveness of treatment, and often transforms an acute and potentially reversible illness or injury into a chronic, irreversible condition that involves permanent disability (Day 2013; Esmail and Globerman 2013). Therefore, waiting

S.A. Rahimi (✉) • A. Jamshidi • D. Ait-Kadi
Department of Mechanical Engineering, Faculty of Science and Engineering, Université Laval, Pavillon Adrien-Pouliot 1065, av. de la Médecine, Local 3517, Québec (Québec), G1V 0A6, Canada
e-mail: samira.abbasgholizadeh-rahimi.1@ulaval.ca

A. Ruiz
Faculty of Business Administration, Université Laval, Pavillon Palasis-Prince 2325, rue de la Terrasse, Local 2423, Québec (Québec), G1V 0A6, Canada

A. Matta et al. (eds.), *Health Care Systems Engineering for Scientists and Practitioners*, Springer Proceedings in Mathematics & Statistics 169,
DOI 10.1007/978-3-319-35132-2_3

lists management and appropriate patients' prioritization can play an important role in diminishing these undesirable outcomes.

Prioritization is a complex multi-criteria decision making process. Scoring systems have been designed as decision making tools to guide surgeons and clinicians. Scoring systems were introduced in 1990s, but afterwards, these initial scoring systems were criticized for being largely arbitrary and resulting in significant numbers of patients being mistakenly denied treatment (with fatal consequences) (Fraser et al. 1993). Afterwards, other scoring systems have been created and used in Italy, Sweden, New Zealand, United Kingdom, Norway, Germany, Canada and other OECD countries. In Canada, the "Western Canada Waiting List" (WCWL) project (Noseworthy et al. 2003) put together several organizations with the aim to develop scoring tools for priority setting in five clinical areas: cataract surgery; general surgery procedures; hip and knee replacement; magnetic resonance imaging scanning, and children's mental health, and since then, the resultant scoring systems have been widely used.

Although these prioritization tools are useful, they still suffer some limitations. In fact, current systems shows inaccuracy in the prioritization process, overlook potential medical risks associated to patients, don't integrate group decision making, and overlook the interdependency among prioritization criteria. There are other major shortcomings in current prioritization systems that need to be improved. These shortcomings are explicitly stated in (Rahimi et al. 2016). In this context, our study focuses on developing prioritization approaches to overcome some of these shortcomings, and rank patients on waiting lists for elective surgical procedures in a more effective and equitable way. To this end, we use Multi-Criteria Decision Making models. Because of the specificities of the problem, both Group Analytic Hierarchy Process (AHP) and Analytic Network Process (ANP) models seem to be attractive alternatives. The use of AHP and ANP also make the model easy to implement, use and update if necessary. The rest of the paper is organized as follows: the next section proposes criteria for surgical patients' prioritization based on the literature. Then, AHP and ANP models for solving the prioritization problem are presented. A numerical example is proposed in section "Models for Prioritization of Patients" to illustrate how these models are applied, and the results produced by both models are. A concluding discussion and future research avenues are provided in the last section.

Choosing Criteria for Surgical Patients' Prioritization

Identification and description of relevant criteria is one of the most important steps in the prioritization procedure. Recently, a list of relevant criteria to assessing patient prioritization was proposed in (Domènech et al. 2013). However, in that work criteria were used in a scoring system where one single decision maker assess patients and uses the provided evaluation scales. In this work, the eight criteria proposed in that work are reconsidered. To better detail the third criterion and introduce the notion of risk, we propose six sub-criteria. Moreover, we use these criteria in a group-decision environment and inside a structured decision making approach which allows capturing relationships between criteria. All the criteria (C1–C8) and subcriteria (C31–C36) are presented in the following:

C1—Disease severity. Severity concerns the consequences of disease on patient's health or in detail on his/her organ function. A precise assessment of disease severity is a rather difficult task, and generally is done upon clinical examination, tests or existent clinical severity scales.

C2—Pain (or other main symptoms). Pain refers to suffering in general or more specifically about physical pain. It may be keen, dull, come and go, or it may be constant. Pain refers to the degree of the main symptom (type, intensity or frequency) affecting daily life activities and health related quality of life (Domènech et al. 2013).

C3—Rate of disease progression. It varies between patients based on many factors like their disease, their health state, age and etc. Earlier prioritization systems used to evaluate this criterion by three absolute choices: *No progression*, *Worsening claudication in the last 3 months*, and *restenosis from previous intervention.* However, we consider this criterion requires a more accurate definition, so we propose six main risks as sub-criteria. These sub criteria are now presented.

C31: Risk of death. It concerns the probability of death for a patient waiting for surgery and it is one of the most important sub-criteria in our model. Since patients waiting for the same surgery may show different levels of risk, this sub-criterion seeks to ensure that patients with higher risk of death will be selected first.

C32: Risk of serious complications, development of comorbidity or worsening the severity of the illness. This subcriterion assesses the probability a waiting patient develops one or more additional diseases (comorbidity) or serious complications on his health state.

C33: Risk of reducing the effectiveness of surgery. Delays can reduce the effectiveness of surgery. For instance, in very old people the effectiveness of joint surgery decreases.

C34: Past complications (R4). Past health situation and complications of patient may impact on surgery outcome and even constitute a threat to patient's health.

C35: Risk of affecting adjacent organs or spread of the disease (R5). This sub-criterion considers the need for fast surgery for patients with high possibility of spread of disease and affecting near organs.

C36: Progression that might affect the survival or can modify the type of surgery (R6). Any kind of delay in surgery which might affect the survival probability of a patient or which could force surgeons to modify the type of surgery to perform.

C4—Difficulty in doing activities. Difficulty in doing activities concerns patient's limitations to perform daily life activities that he/she was able to do prior

Table 1 Criteria, and weights for elective surgery (Domènech et al. 2013)

C1	C2	C3	C4	C5	C6	C7	C8
0.23	0.14	0.15	0.14	0.12	0.05	0.08	0.09

to his/her disease. Although there is a relationship between this criterion and Severity, they should not be confused (Domènech et al. 2013).

C5—Probability and degree of improvement. It assesses the overall improvement in health related quality of life (Domènech et al. 2013). Since this criterion estimates expected benefit of surgery it is considered in our study as a benefit related criterion.

C6—Time on waiting list. The time spent on the waiting list is one of the most important factors from patients' point of view and thus affects the patients' perception on the quality of the received services.

C7—Limitation to care for one's dependents. It concerns limitations to exercise the responsibility of taking care of dependents (i.e. children, elder parents) (Domènech et al. 2013).

C8—Limitations in the ability to work/study/seek employment. This criterion indicate limitation to work (in paid or unpaid jobs) because of the condition or limitation for schooling or educational activities and job-seeking (Domènech et al. 2013). Criteria 7 and 8 highlight the social role of patients in the prioritization problem.

Domènech et al. (2013) weighted these criteria in collaboration with panel of medical experts. Then, in the second round, they included other stakeholders (they asked 112 experts to give weights to these 8 criteria). Their response rate was 52 %. Table 1 reports the weights assigned to each criterion based on (Domènech et al. 2013). If a new criterion is added or an existing one is deleted from the model, all the criteria's weights must be reassessed.

Models for Prioritization of Patients

This part introduces AHP and ANP, two multi-criteria decision making models, and presents how they are used to help prioritizating surgical patients.

Prioritization Model Using Analytic Hierarchy Process (AHP)

Analytic Hierarchy Process (AHP) is a decision making tool which has been successfully applied to complex situations where several decision makers deal with multiple factors having diverse relative importance. AHP decomposes a problem into a hierarchy in which the overall decision objective lies at the top and the criteria, sub-criteria and decision alternatives are on each descending level of the hierarchy (Partovi et al. 1989). Decision makers compare each factor to all the other ones in

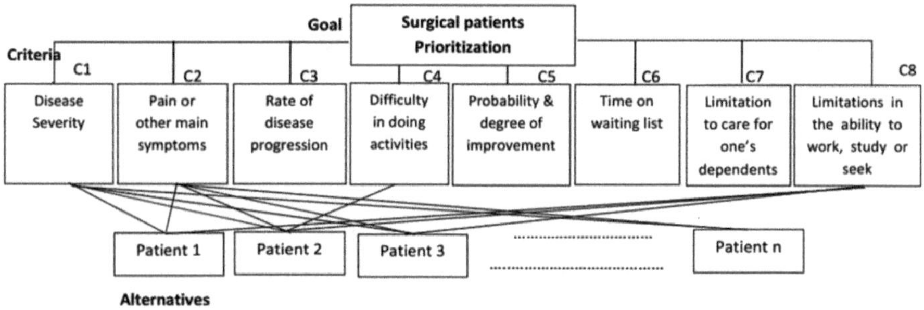

Fig. 1 Hierarchy model for prioritization of surgical patients

the same level of the hierarchy using a pairwise comparison matrix to find each factor's weight or relative importance. The "best" solution is the alternative, i.e. the patient, having the greatest cumulative weight (Saaty 1990a). We refer the interested reader to (Saaty 1990b) for a thorough discussion on AHP.

Our first model considers prioritization of surgical patients as a MCDM problem and use group AHP to solve it. In this part we explain the proposed hierarchy model to solve this critical problem. The development of the hierarchy is the first fundamental step to explain the problem structure more clearly and specifically. To do so, all the important criteria must be identified and placed at the appropriate level of the hierarchy. In the previous section we elected eight main criteria from the literature. Patients compose the third level of the hierarchy. The objective is to assign a criticality score (or priority) for every single patient participating in the model. Figure 1 shows a decision hierarchy for prioritization of surgical patients.

To assess a patient with respect to a particular criterion, the criterion's description and their associated intensities should be defined in advance. After assigning weights to all criteria and medical experts (e.g. surgeons and/or clinical staff) the model is ready to be used to assess the patients' priorities. A numerical experiment illustrating this model is provided in Sect. 4.

Prioritization Model Using Analytic Network Process (ANP)

"Many decision problems cannot be structured hierarchically because they involve the interaction and dependence of higher level elements in a hierarchy on lower level elements" (Saaty and Özdemir 2005). Since AHP allows just one-directional hierarchical relationships, more flexible tools able to model complex interrelationships among factors and decision levels are needed. The Analytic Network Process (ANP) was developed by (Saaty 1990, 2008) to overcome the problem of dependence and feedback among criteria in AHP. Such "complex interrelationships among criteria" seem to be present in our context, so we remodel our prioritization problem as a network (ANP) model in order to compare the results produced by both approaches. Briefly, ANP encompasses the following four phases:

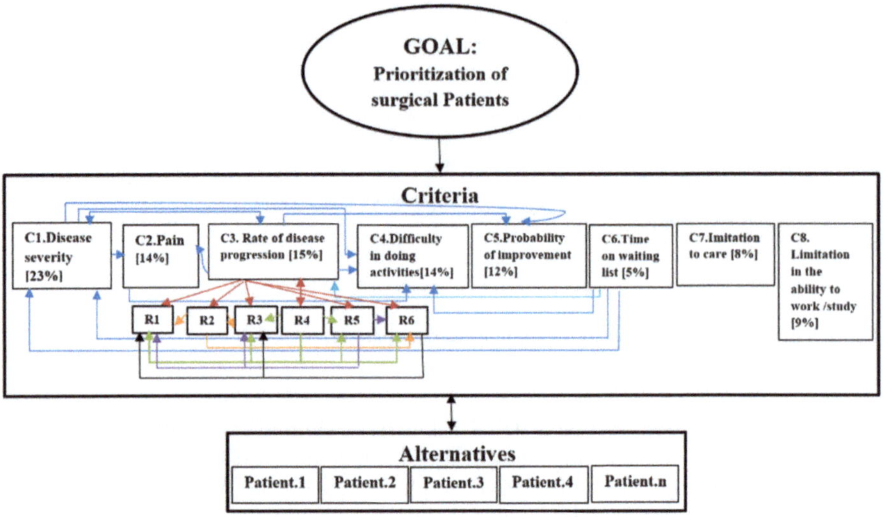

Fig. 2 Decision network for surgical patients' prioritization patients

Phase 1: Construction of model and problem structuring; the prioritization problem is stated clearly and decomposed into a rational network (see Fig. 2). The ANP technique builds a network model of the problem by structuring it into clusters containing elements that are related to, or influence each other. The medical experts determine the influence relationships between model elements based on their knowledge and their experience. A brainstorming or other appropriate discussion method can be used in this step.

Phase 2: Pairwise comparisons and priority vectors construction; Like in the case of AHP, pairs of decision elements at each component/cluster are compared with respect to their importance towards their control criterion, and the components themselves are also pairwise compared with respect to their contribution to the main goal (surgical patients' prioritization). In addition, interdependencies among criteria of a cluster must also be examined in pairs. Medical experts are asked to respond to a series of pair-wise comparisons (two elements or two components at a time will be compared). The relative importance values are determined by using Saaty's 1-to-9 scale [15], where a score of 1 stand for equal importance between two factors and a score of 9 presents the extreme importance of one factor compared to the other one.

Phase 3: Formation of Supermatrix; the influence of elements on other elements in the network is represented by the "supermatrix" (Saaty 1990a; 2005; 2008). To obtain global priorities in a system with interdependent influences, the local priority vectors are entered in the appropriate columns of a matrix. Therefore, a "supermatrix" is a partitioned matrix where each partition represents a relationship between two nodes (components/clusters) in a system. The elements in the matrix are obtained by normalizing each block of the supermatrix.

Phase 4: Synthesis of the criteria and patients' priorities and selection of the best patient; the priority weights of patients are found in the column of alternatives in the normalized supermatrix. The patient with the largest overall priority should be the first to be selected.

Numerical Application

We propose a numerical example to illustrate the application of the proposed models and to compare their performance. The example concerns four patients with different profiles and medical states. We assume that three experts (clinical staff and/or surgeons) assign to each patient a value in the 1–9 scale to each considered criterion. Experts receive also weights based on their knowledge and experience. In our example, we set experts' weights to 0.126, 0.168, and 0.706 for expert 1–3, respectively. The goal of both models is to prioritize surgical patients. To this end, both models incorporate eight main criteria, and in the ANP model, six additional sub-criteria are considered. The four patients are considered as "alternatives" in both models. Related hierarchy and network structures are shown in Figs. 1 and 3, respectively.

The overall priorities of the patients produced by the group AHP and ANP models are shown in Table 2. The numerical results were computed by *Super Decisions* software version 2.0.8 in the case of ANP, and *Expert Choice*, for AHP. As it can be observed, both models lead to similar conclusions for the highest rank patients P1 and P3. However, such a small experiment does not allow us to compare the performance of the proposed models nor to elect one rather than the other.

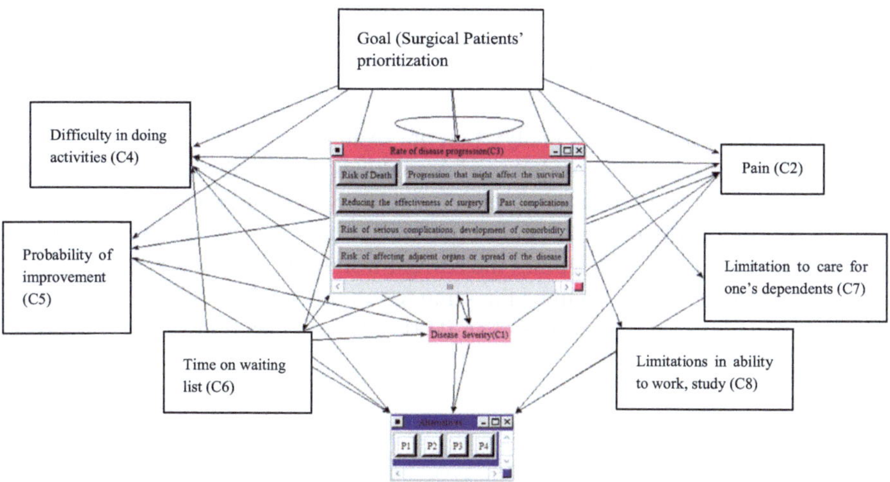

Fig. 3 network structure of surgical patients' prioritization

Table 2 Results produced by AHP and ANP models

Patients	Score AHP	Rank AHP	Score ANP	Rank ANP
P1	0.425	1	0.4722	1
P2	0.131	4	0.2000	3
P3	0.238	2	0.2293	2
P4	0.206	3	0.0985	4

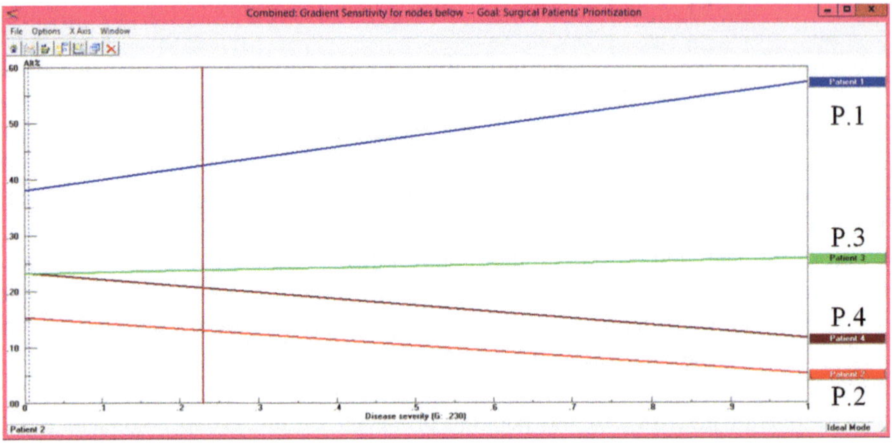

Fig. 4 Sensitivity analysis with respect to criterion C1

Sensitivity analysis is used to assess the model reliability in response to changes on the weights assigned to considered criteria. Sensitivity is important because users always wonder about how a slight different choice of weights could have led to very different results. In a sensitivity analysis, the weights assigned to criteria are modified and the system responses are recorded and analyzed. Ideally, the model's outcome, (i.e., the ranking of the patients), should be stable (unchanged) under minor variations of weights values.

We conducted a sensitivity analysis on our example with four patients. Figure 4 graphically illustrates the changes in the rankings for all the patients when the weight for criteria "C1" (disease severity) is changed, the weighted values for other criteria are equally distributed as the weight of C1 changes. The solid vertical line at the weight of 0.23 in Fig. 4 denotes the current weight for disease severity. Figure 4 also shows that the ranking will be the same for all the possible values of C1 greater than 0.006. If C1 is weighted less than 0.006, P4 becomes preferred with respect to P3.

Discussion and Future Work

This paper addressed the difficult problem of prioritizing patients in surgical waiting lists and presented two multi-criteria decision making models in an attempt to overcome the limitations of the currently used prioritization methods. After reviewing the pertinent literature, eight criteria and six sub-criteria (as associated risks) were elected to assess the relative priority of patients on surgical waiting lists. Criteria were balanced and structured by means of Group Analytic Hierarchy Process (AHP) and Analytic Network Process (ANP), two well-known multi-criteria approaches to determine weighting values for the criteria and their grades' intensities, and to evaluate the surgical patients. Patients with lower criticality scores can be assigned a lower priority in scheduling program. We illustrated the potential of our models on an academic small sized instance. Despite of the limited scope of our experiments, the results suggested that the proposed models not only consider various perspectives determining patients' priorities, but also remain noticeably stable and robust as shown in sensitivity analysis.

Although we believe that the proposed models are efficient and effective, there are areas for future enhancements. Future research should extend the current models and use the fuzzy decision variables in place of single numeric values in order to incorporate these inherent uncertainties. Fuzzy set theory may reduce the vagueness associated with decision makers.

Finally, this research can be linked to the one concerning operational surgical activities scheduling. In fact, the real question addressed by operation theaters' managers is how to use the available resources in order to serve the patients who need it the most. This research clearly contributes to identify those patients.

Acknowledgements This research was partially financed by discovery grants [OPG 0293307 and OPG 0118062] from the Canadian Natural Sciences and Engineering Research Council (NSERC). This support is gratefully acknowledged.

References

Abbasgholizadeh Rahimi, S., Jamshidi, A., Ruiz, A., Ait-Kadi, D.: Applied methods in prioritization of patients in surgery waiting lists. In: Proceedings of the Industrial and Systems Engineering Research Conference, Montreal, Canada, 1–3 June 2014

Abbasgholizadeh Rahimi, S., Jamshidi, A., Ruiz, A., Ait-Kadi, D.: A new dynamic integrated framework for Surgical patients' prioritization considering risks and uncertainties, to be published in Journal of Decision support systems (2016)

Barua, B., Esmail, N.: Why Are We Waiting So Long? Health Care Wait Times Nearly Double over the Past Two Decades to 18.2 Weeks. Fraser Institute, Canada (2013)

Day, B.: Reducing Wait Times for Health Care. Fraser Institute, Canada (2013)

Domènech, M.S., Adam, P., Tebé, C., Espallargues, M.: Developing a universal tool for the prioritization of patients waiting for elective surgery. J. Health Policy **113**, 118–26 (2013)

Esmail, N., Globerman, S.: Canada's Long Wait Times for Health Care Leading to Patient Depression and Death. Fraser Institute, Canada (2013)

Fraser, G., Alley, A., Morris, R.: Waiting lists and waiting times: their nature and management: a report to the National Advisory Committee on Core Health and Disability Support Services. (1993)

Mullen, P.M.: Prioritizing waiting lists: how and why? EJOR **150**, 32–45 (2003)

Noseworthy, T.W., McGurran, J.J., Hadorn, D.C.: Waiting for scheduled services in Canada: development of priority-setting scoring systems. J. Eval. Clin. Pract. **9**, 23–31 (2003)

Partovi, F.Y., Burton, J., Banerjee, A.: Application of analytic hierarchy process in operations management. Int. J. Oper. Prod. Manag. **10**, 5–19 (1989)

Randolph, H.: Patient flow reducing delay in healthcare delivery. In: International Series in OR & Management Science, vol. 206, 2nd ed., p. 107, Springer US (2013)

Saaty, T.L.: How to make a decision: the ANP. EJOR **4**, 9–26 (1990a)

Saaty, T.: The Analytic Hierarchy Process. RWS Publications, Pittsburgh PA (1990b)

Saaty, T.L.: Decision making with the ANP. Int. J. Serv. Sci. **1**, 83–98 (2008)

Saaty, T.L., Özdemir, M.S.: The Encyclion: A Dictionary of Decisions with Dependence and Feedback Based on the ANP. RWS Publications, USA (2005)

US National Library of Medicine. Medical subject headings.http://www.nlm.nih.gov/cgi/mesh/2012/MBcgi (2012). Accessed 19 July 2012

Bed Managers: The Patient's Personal Assistant

Seren Schirra, Gaelle Olleon, Estelle Forestier, Sylvie Meyran, Emmanuel Beaudry, and Marie Lassaigne

Introduction

Emergency departments in France are often saturated, "victims" of a vast number of patients seeking medical care, and of logistical problems linked to the lack of beds for patients waiting to be admitted (Green and Armstrong 1993, 1995). For over 10 years now, our ward has emergency nurses specially designated to facilitate patient orientation after emergency care. Over the years, the role and responsibilities of these nurses has evolved and improved. Their expertise is recognized at the highest level of the hospital administration. More recently in other emergency wards it has also become common practice, strongly backed by the Ministry of Health. These nurses are commonly known as Bed Managers. We would like to share our experience with you.

General Presentation of the St Luc St Joseph Hospital

The St Luc St Joseph hospital is a private structure, but open to general public and makes no personal profit.

The hospital employs over 1100 people, including over 200 doctors and 90 medical students. It is a relatively new building, inaugurated in 2002, although historically its origins date as far back as 1869.

Its city center location makes it easily accessible both on foot and by public transport.

S. Schirra • G. Olleon • E. Forestier • S. Meyran (✉) • E. Beaudry • M. Lassaigne
Centre Hospitalier St-Joseph/St-Luc, Lyon, France
e-mail: smeyran@ch-stjoseph-stluc-lyon.fr

A. Matta et al. (eds.), *Health Care Systems Engineering for Scientists and Practitioners*, Springer Proceedings in Mathematics & Statistics 169,
DOI 10.1007/978-3-319-35132-2_4

The population treated is varied: the local "lyonnais" population tips slightly towards elderly, but the region has a strong working population as well, largely influenced by the presence of the industrial zone offering jobs in pharmaceuticals, biology and medical research and petro-chemicals. Set in the heart of Europe, Interpol has offices here, and various universities and schools bring a younger generation of students to the city. The rich cultural heritage (music festivals, cinema festival, the Lights festival, food festivals . . .) and its reputation as food capital of the world make Lyon an appealing destination for many international visitors.

What the Hospital Offers in Terms of Care

The hospital is what the French refer to as an MCO (medicine, chirurgic, obstetric) and has 350 beds. One of our distinguishing characteristics is that the wards are not specialty specific; meaning that on two floors the various different medical specialties are treated on a same ward (over 150 beds), and one floor is designated to surgical patients (100 beds). We also provide a day hospital for both medical and surgical patients, a ward in considerable expansion over the past few years. The hospital has a level 1 maternity ward (available only to low risk pregnancies and births), an intensive care unit and a burns unit. The technical platform is comprised of 12 operating theatres (8200 surgeries a year), a cardiology and rhythmology unit, a medical imagery unit (scan, MRI, X-ray) a biology lab and a dialysis ward.

The Emergency Department (ED)

The Emergency department comprise four wards:

- The SMU or emergency room, located on the ground floor, with its own individual access for pedestrians and medical vehicles.
- The UHCD or post emergency room ward located on the first floor and has 13 beds. The average length of stay is between 24 and 48 h.
- The UHP or varied treatment ward, located on the third floor and has 20 beds, the average stay is 48 h or more.
- The SST or "follow up" trauma center for patients initially treated on the emergency ward and needing follow up care.

The ward provides medical care for adults only, a few minor pediatric cases can receive initial care before being transferred to the pediatric hospital of Lyon.

The Emergency Ward Team

The emergency ward team is multidisciplinary; made up of 130 people—doctors, nurses, administrators. Fifteen to twenty-four people are present 24 h a day, 7 days a week all year round. They provide care and assistance for the daily average of 100 patients seeking medical attention. Twenty percent will need to be admitted. The organization of their admittance is handled by the Bed Managers.

The Organization of Emergency Patient Admittance

Inter Hospital Admittance

Taking into account all the patients admitted to St Joseph St Luc hospital via the emergency ward, 65 % of them will remain within the emergency department, either on the post emergency ward or on the varied treatment ward. The remaining 35 % will be transferred to the medical and surgical wards.

Admittance to Different Hospitals

Transfer to a different hospital for admittance happens for various reasons, the main reasons being:

- The patient requires care in a specialty that St Joseph St Luc does not provide.
- The patient is already being treated by a doctor or specialist in another hospital.
- When the patient influx is higher than the admittance capacity of the hospital.

Bed Managing: Or How to Coordinate the Patient Influx and Optimize Care

Within the Hospital in General: The Central Reservation Team

The mixed specialty and surgical wards require the presence of a designated admittance team for bed management. This refers to a team of two nurses and two secretaries who are in contact with every ward within the hospital and via who any requests for admittance must pass, be they internal or external.

The team plans days ahead at a time, and dispatches the available beds taking into account the requirements of all those concerned (Gossart et al. 2010).

Within the Emergency Department: The Bed Managers

Patients who arrive via the emergency ward and who need to be admitted are put into the care of a team of highly experienced emergency nurses who we refer to here as Bed Managers (Proudlove et al. 2007). They work alongside the geriatrics team, the psych team, the ward nurses and doctors and of course the central reservation team.

These nurses organize and coordinate patient care from the moment the doctor asks for admittance until the moment the patient is in their designated bed.

The emergency ward is the only ward in the hospital to benefit from a bed managing team, and was one of the first of its kind, having created the job almost 15 years ago.

The Bed Managers

Job Profile

- Experience of work on the emergency ward (in depth knowledge of the medical staff on the ward and in the hospital, knowledge of the type of patients treated and the workload).
- In depth knowledge of the region (the different medical structures and their location).
- A dynamic and sociable personality.
- An acute sense of diplomacy and tact.
- Managerial and pedagogical skills.

Tools

- Clinical sense- "nurses' instinct" (aided by experience working on the ward).
- Extensive and detailed knowledge of the patient and their medical file.
- An extensive address book and personal network.
- A telephone!

Key Points

Bed Manager: A Personal Assistant

- They are on site from 8 to 18 h, 7 days a week, all year round.
- They are on the ward all day, most importantly for the medical shift change twice a day. This gives them visibility on patient influx and workload, vital in order to anticipate admittance.

- The bed managers use their clinical common sense to prioritize the patient evaluation by the doctors, and are deeply involved in the patients' healthcare journey.
- The bed managers give an exact and detailed account of the patients to other healthcare professionals, particularly in the event of a transfer outside the hospital.
- They anticipate patients discharge.
- The bed managers organize and anticipate the various tests and consultations necessary for the patient, always aiming to reduce waiting time and bed blocking.

Bed Manager: An Intermediary

- The main preoccupation of the bed managers is to see the patient in their globality, taking into account every aspect of the person.
- They are a direct link between the patient, their family, the medical team and the other wards.
- The bed managers coordinate the patients' healthcare journey, alongside the central reservation team.
- They are the trusted spokespeople for the patients and doctors when presenting a case to another specialist or hospital.
- They are the central pivot of the ward, coordinating between the front and back office, between nurses and doctors, patients and staff.
- The bed managers are highly respected by their peers.

Bed Manager: A Mediator

- They aim to improve the quality of care.
- The bed managers help to improve the working conditions of the rest of the team, often resolving minor logistical problems or conflicts.
- They have an expert knowledge of the medical equipment within the ward.
- In the event of a massive influx, the bed managers resume their initial function of nurse, to aide and support the team.
- In the absence of the head nurse, the bed managers take up the managerial position, for example on weekends/bank holidays.
- They work hand in hand with the head nurse in the evaluation of new members of the nursing team.
- They use their expert knowledge and experience to help train the nursing students, and act as their "go – to" person throughout their work placement.
- The bed managers are spokespeople for the ward with regard to the hospital administrators.

Bed Manager: A Few Minor Setbacks

As with any job there are also more negative aspects. In this case it is clear that the negative aspects of bed managing can be highly detrimental to patient welfare and more precisely with regards to waiting time in A&E, admittance after A&E or transfers elsewhere.

One of the main difficulties faced on a daily basis is the sheer number of people involved in the care of one single patient. Coordinating and relaying the information between the various health care professionals can be difficult, even more so when relying on people outside of the ward to give the green light (radiologists, operating theatre, ward sister, ambulances . . .).

It is also complicated trying to transmit the information by telephone whilst simultaneously being given more information from someone else!

Bed managers hold a pivotal place in this emergency ward, but this is also sometimes a source of tension within the team; bed managers are neither ward sisters nor general nurses, they are esteemed by the doctors, and yet no longer have a place amongst the nursing team. Their peers often "forget" that the bed managers may be called upon for technical support as well.

With the expert knowledge the bed managers have of the ward comes the inevitable idea that they know "everything" and can solve any problem brought to them. Whilst this is flattering and almost true, it does mean that a considerable amount of time is lost solving minor issues such as a broken printer, a lost item of clothing for a patient, a dispute amongst team members.

Another major difficulty they are faced with daily is quite simply the "habits" of co-workers and doctors within the hospital: some specialists wont speak to a nurse when a consult is needed, some prefer to have a request faxed. Some nursing teams wont admit patients over the lunch break or after 17h30, some wards refuse to do blood transfusions on the night shift.

The list is long and all these minor details contribute to making the bed managers extremely efficient negotiators but also make their job a lot harder.

Conclusion

Bed management, as put forward by the Ministry of Health, aims to optimize hospital efficiency, mainly through bed occupation, and reduced waiting times in emergency wards. The pivotal role of the bed managers with regards to this goal does not however change their main priority—patient welfare.

References

Gossart, D., Meskens, N., Guinet, A.: Strategic and tactical hospital bed management in Belgium and in France. In: Camarinha-Matos, L.M., Boucher, X., Afsarmanesh, H. (eds.) Collaborative networks for a sustainable world, vol. 336, pp. 243–250. Springer, Berlin (2010). IFIP Advances in Information and Communication Technology. ISBN 978–3-642-15960-2

Green, J., Armstrong, D.: Controlling the 'bed state': negotiating hospital organization. Sociol. Health Illn. **15**, 3 (1993)

Green, J., Armstrong, D.: Achieving rational management: bed managers and the crisis in emergency admissions. Sociol. Rev. **43**, 743–764 (1995)

Proudlove, N., Boaden, R., Jorgensen, J.: Developing bed managers: the why and the how. J. Nurs. Manag. **15**, 34–42 (2007)

An Optimization Model for Sequence Dependent Parallel Operating Room Scheduling

Johan Holmgren and Marie Persson

Introduction

This paper presents a novel approach for increasing the operating room efficiency by significantly reducing the turnover time in elective surgical case scheduling. The operating room turnover time represents lost revenue-generating opportunity for the hospital and is hence important to control. Reduced turnover time typically leads to increased surgeon utilization and also increased patient throughput (see section "Overlapping Surgical Activities" for further elaboration). In order to reduce the turnover time, some hospitals use overlapping surgical activities, i.e., one surgical case can be initiated before the ongoing surgical case is closed (see, e.g., Friedman et al. 2006).

However, during the overlaps, additional operating room resources are required. To meet the problem of operating room scheduling with allowed overlaps, we present an optimization model that sequences surgical cases and their associated overlaps, subject to a limited number of operating room resources.

Here, one operating room resource is equivalent to one surgical team (consisting of at least: one anesthesia nurse and one surgical assistant nurse). A surgical case consists of several subsequent activities, here simplified into: (i) surgical *pre-procedure*, (ii) surgical *procedure* and (iii) surgical *post-procedure*. We assume that

J. Holmgren
Department of Computer Science, Faculty of Technology and Society, Malmö University, SE-205 06 Malmö, Sweden
e-mail: johan.holmgren@mah.se

M. Persson (✉)
Department of Computer Science and Engineering, Blekinge Institute of Technology, SE-371 79 Karlskrona, Sweden
e-mail: marie.persson@bth.se

A. Matta et al. (eds.), *Health Care Systems Engineering for Scientists and Practitioners*, Springer Proceedings in Mathematics & Statistics 169,
DOI 10.1007/978-3-319-35132-2_5

the surgical pre-procedure may be conducted in a preparation room, a room close to, or somehow connected to, the actual operating room. This means that: for each operating room, the surgical pre-procedure for the upcoming surgery may start before the foregoing surgical case is closed. However, this requires that at least one surgical team is available to undertake the surgical pre-procedure as another surgical team is already occupied closing the ongoing surgical case. After the ongoing surgical case is closed and the surgical pre-procedure for the upcoming surgery is completed, one of the two surgical teams is released. We formulate a mixed integer linear problem (MILP) to model the characteristics of the described problem and conduct an experiment to demonstrate its advantages.

The research problem of operating room planning and scheduling of surgical cases has been extensively studied by quantitative modeling approaches during the last decades (Cardoen et al. 2010; Hans et al. 2008; Testi et al. 2007). Cardoen et al. (2010) present a literature review that highlights the trend in elective case scheduling. They emphasize the steady amount of elective studies using approaches, such as mathematical programming and simulation, for problem solving and classifies the included papers mainly according to technical approaches. In another study by Abdelrasol et al. (2013) the operating room problem is divided according to strategic, tactical and operational level. They define the strategic level to determine the number of operating rooms dedicated to each surgical specialty covering a time period of several months. In the second level they describe the problem of allocating operating room time to each surgeon/surgeon groups within a certain specialty on a weekly basis, while the last level concerns how to select and sequence patients to each operating room. A common policy for elective surgery management on a tactical level is to use the block system. A block is a dedicated time interval with a typical duration of a half or a full day. It concerns one or more operating rooms, which are allocated to a specific surgeon or a surgical group (Guptaa and Denton 2008). Typically, each surgeon, or surgeon group, separately manages their own waiting list, including how to sequence the elected surgeries. Azari-Rad et al. (2013) present a surgery sequencing strategy, where surgical cases are ordered by patient length of stay, and surgery duration and variance. Their aim is to reduce the number of cancellations, whereas we are focusing on achieving increased patient throughput by scheduling patients in such sequence that minimize the turnover time.

In another study, presented by Arnaout and Kulbashian (2008), they schedule each parallel operating room, given sequence dependent setup with the aim of minimizing the makespan. Similarly to Arnout and Kulbashian we present a model that addresses the sequencing problem in parallel operating rooms. However, in our model we allow for partially overlapping surgical activities to maximize the utilization of the operating room.

Our modeling approach is also somehow similar to how Pham and Klinkert (2006) develop a generalized job-shop scheduling model. However, they study the entire patient process from pre-operative holding unit to post-operative care unit. We on the other hand, solely focus on the process of operating room planning.

The rest of the paper is organized as follows. In section "Overlapping Surgical Activities", we discuss the problem of surgical case scheduling with overlapping surgical activities. In section “Optimization Model”, we present our optimization model, followed in section “Computational Experiment” by a presentation of our computational experiment. Finally, we conclude the paper in section “Concluding Remarks”.

Overlapping Surgical Activities

In Fig. 1, we demonstrate how a surgical case can be divided into three subsequent activities as aforementioned (surgical pre-procedure, surgical procedure and surgical post-procedure). During the surgical pre-procedure, the anesthesia is started and the surgical instruments are prepared. During the surgical procedure, the actual surgical operation is conducted, whereas during the surgical post-procedure, the patient is stitched up, the anesthesia is stopped, the instruments are wrapped up, and the room is cleaned up. It should be noted that the surgical post-procedure is not equivalent to post-operative care. Post-operative care is conducted by other staff and commonly takes place in another room (possible in another department), in which several patients are monitored simultaneously.

Consider the situation where we have two surgical cases, A and B, which are planned in the same operating room (see Fig. 2). During the surgical post-procedure of surgery A, we hypothesize that the surgical pre-procedure of surgery B can be conducted in the preparation room and continued in the operating room as soon as the surgical post-procedure of surgery A is completed. This leads to an increased utilization of the surgeon since the surgeon is not forced to wait for the surgical pre-procedure of the upcoming surgical case is completed. Instead the surgeon can start the upcoming surgery almost immediately after the ongoing surgical case (surgery A) is completed. In this way, time will be saved and the number of surgery, patient throughput, can be increased. The surgical post-procedure includes cleaning the operating room, which is mandatory between surgeries and hence, the surgical procedure of surgery B is not permitted to start before the surgical post-procedure of surgery A is completed. As mentioned above, there must be at least two available surgical teams during this overlap.

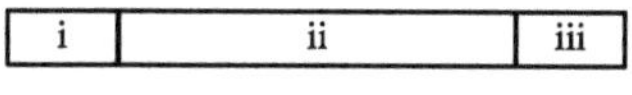

Fig. 1 A surgical case with (i) surgical pre-procedure, (ii) surgical procedure, and (iii) surgical post-procedure

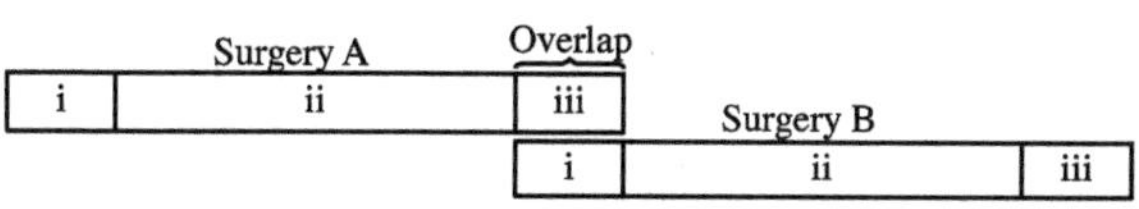

Fig. 2 Two subsequent surgical cases, A and B, in an operating room, where the surgical post-procedure of A overlaps with the surgical pre-procedure of B

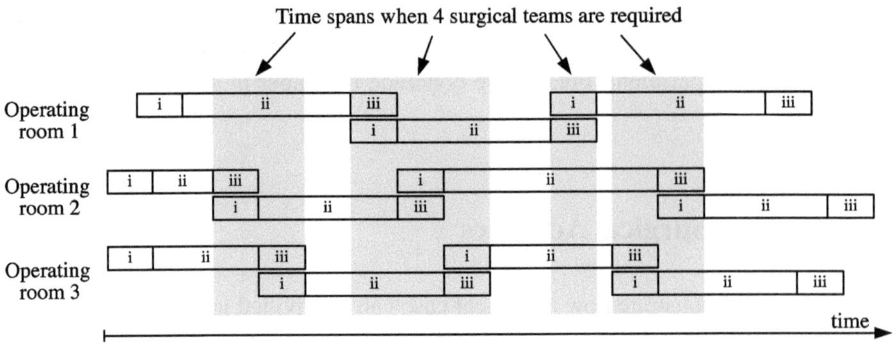

Fig. 3 A schedule for a surgery department with three operating rooms and four surgical teams

In Fig. 3, we provide a possible surgery schedule for a surgery department consisting of three operating rooms and four surgical teams. The challenge is to schedule surgical cases in order to make maximum use of the overlaps subject to a limited number of surgical teams.

Optimization Model

In this section, we present a mixed integer linear optimization problem (MILP), which captures the main characteristics of the considered problem. The objective of the optimization model is to identify a sequence of surgical cases during a day in such a way that the need for overtime of the surgical teams is minimized.

We let T denote the opening hours for a day (specified in minutes), O denote the set of operating rooms, and S_o denote the set of surgical cases that have been assigned to operating room $o \in O$ during a specific day ($S = \bigcup_{o \in O} S_o$ is the set of all surgical cases). For surgical case $s \in S$, t_s^{pre} is the expected surgical pre-procedure time, t_s^{surg} is the expected surgical procedure time, and t_s^{post} is the expected surgical post-procedure time. In addition, t^{prep} is the expected time needed for preparing (primarily cleaning) the operation room for a surgical case. This time is assumed to be identical for all operating rooms and for all surgical cases. Finally, r is the number of available resources, i.e., the surgical teams. Each surgical case demands one surgical team to be present during each of the three surgical activities (pre-procedure, procedure, and post-procedure); however, all of the three activities do not necessarily need to be assigned to the same team. It should be noted that the surgeon is typically only present during the actual surgical procedure. In the model, we assume that there is one or more surgeons assigned to each operating room, the exact number depends on the surgical case, and we therefore do not explicitly include the scheduling of surgeons in the optimization model.

For each surgical case $s \in S$, we use a set of decision variables $x_s^{pre\ start}$, $x_s^{surg\ start}$, $x_s^{post\ end} \in \mathbb{Z}^+ \cup \{0\}$ in order to determine the surgical pre-procedure and procedure start times, and the surgical post-procedure end time of s. We do not explicitly model the surgical post-procedure start time, because we assume that both the surgical procedure and surgical post-procedure of a surgical case is assigned to the same surgical team. In addition, the surgical pre-procedure end time, the surgical procedure end time, and the surgical post-procedure start time are implicitly given by $x_s^{pre\ start}$ and $x_s^{surg\ start}$, since we assume that there is no time between two subsequent activities for a surgical case. In addition, we introduce variable v_o in order to model the expected overtime in operating room $o \in O$.

An important part of the optimization model is to guarantee that the number of parallel activities (i.e., surgical pre-procedure, surgical procedure, and surgical post-procedure) never exceeds the number of available surgical teams. In particular, the surgical pre-procedure of surgical case $s \in S$ is allowed to start at a particular time only if there are at most $r - 1$ other ongoing surgical activities at that specific point in time. For each surgical case $s' \in S$, it is therefore necessary to capture which other surgical cases $s \in S \setminus \{s'\}$ are ongoing at the surgical pre-procedure start time of s'.

In Fig. 4, we present the 13 possible cases for how two surgical cases (s' and s'') can relate to each other in time. It could be argued that case 1 is equivalent to case 2 as none of them involve overlap, and that case 9 is equivalent to case 10; however, for completeness we have chosen to include all of these four cases in the description of our model. In the figure, it can be seen that s'' is ongoing at the surgical pre-procedure start time of s' in case 6, 7, 8, 11, 12, and 13, and s' is ongoing at the surgical pre-procedure start time of s'' in case 3, 4, 5, 6, 7, and 8.

In order to model which of the other surgical cases that are ongoing at the start of a specific surgical case, we introduced two sets of binary variables: $y_{s's''}, s', s'' \in S, s' \neq s''$ and $z_{s's''}, s', s'' \in S, s' \neq s''$. For two surgical cases $s', s'' \in S, s' \neq s''$, the two variables $y_{s's''}$ and $z_{s's''}$ are used together in order to determine if s'' is ongoing at the start of s'. The purpose of $y_{s's''}$ is to identify whether the surgical pre-procedure of s'' starts earlier than, or at the same time as, the surgical pre-procedure start time

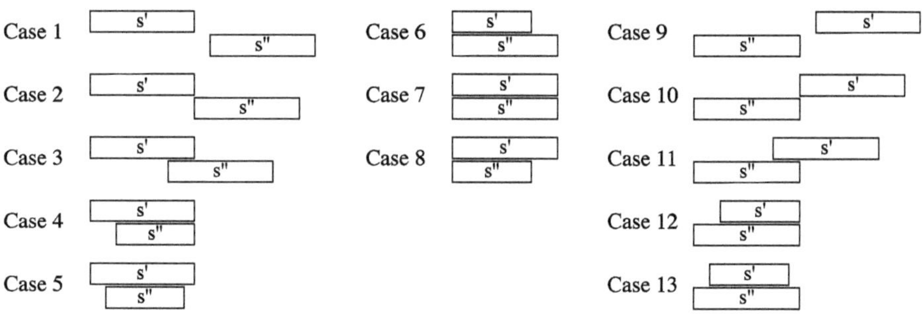

Fig. 4 The 13 possibility ways that two surgical cases (including surgical pre-procedure and post-procedure) can relate to each other in time

Table 1 The values of the y and z variables for each of the 13 possibilities for how two operations s' and s'' can relate to each other in time

Case	$y_{s's''}$	$z_{s's''}$	$y_{s's''} - z_{s's''}$	$y_{s''s'}$	$z_{s''s'}$	$y_{s''s'} - z_{s''s'}$
1,2	0	0	0	1	1	0
3,4,5	0	0	0	1	0	1
6,7,8	1	0	1	1	0	1
9,10	1	1	0	0	0	0
11, 12, 13	1	0	1	0	0	0

of s'. We let $y_{s's''} = 1$ if $x_{s'}^{pre\ start} \geq x_{s''}^{pre\ start}$ and $y_{s's''} = 0$ if $x_{s'}^{pre\ start} < x_{s''}^{pre\ start}$. Each of the eight cases 6–13 correspond to a situation where $y_{s's''} = 1$. In particular, these eight cases include all situations where s'' is ongoing at the start of s', i.e., case 6, 7, 8, 11, 12, and 13. However, they also include two cases where s'' is not ongoing at the start time of s', i.e., where s'' has both started and ended before the surgical pre-procedure start time of s', i.e., cases 9 and 10; the purpose of $z_{s's''}$ is to identify these two cases. We let $z_{s's''} = 1$ if $x_{s'}^{pre\ start} \geq x_{s''}^{post\ end}$ and $z_{s's''} = 0$ if $x_{s'}^{pre\ start} < x_{s''}^{post\ end}$.

The $y_{s's''}$ and $z_{s's''}$ variables can then be combined in order to determine whether s'' is ongoing at the start time of s': surgical case s'' is ongoing at the start time of s' iff $y_{s's''} - z_{s's''} = 1$, and s' is ongoing at the start time of s'' iff $y_{s''s'} - z_{s''s'} = 1$.

In Table 1, it can be seen in the column $y_{s's''} - z_{s's''}$ that case 6, 7, 8, 11, 12, and 13 are those where s'' is ongoing at the start of s', and the column $y_{s''s'} - z_{s''s'}$ shows that case 3, 4, 5, 6, 7, and 8 are those where s' is ongoing at the start of s''.

The objective of the optimization model is to minimize the sum of the overtime cost over all of the operating rooms $o \in O$, i.e.,

$$\sum_{o \in O} v_o \cdot p, \tag{1}$$

where p denotes the overtime cost.

For each surgical case $s \in S$, the relations between the surgical pre-procedure start time, the surgical procedure start time, and the surgical post-procedure end time, are modeled using the constraint sets

$$x_s^{surg\ start} = x_s^{pre\ start} + t_s^{pre}, s \in S, \text{and} \tag{2}$$

$$x_s^{post\ end} = x_s^{surg\ start} + t_s^{surg} + t_s^{post}, s \in S. \tag{3}$$

The overtime (v_o) for each of the operating rooms $o \in O$ is determined using the constraints

$$v_o \geq x_s^{post\ end} - T - 1, s \in S_o, o \in O. \tag{4}$$

The reason for adding '-1' on the right hand side is that we use a time representation where activities are assumed to both start and end in the beginning of periods. For example, if an activity starts in minute t' (actually in the beginning of t'), and is expected to run in t'' minutes, then the activity is expected to end in the beginning of minute $t' + t''$. The next activity in the same room is allowed to start in minute $t' + t''$.

The values of the y and z variables are determined using the constraint sets

$$x_{s'}^{pre\ start} - x_{s''}^{pre\ start} + 0.5 \leq y_{s's''} \cdot M, s', s'' \in S, s' \neq s'', \quad (5)$$

$$x_{s'}^{pre\ start} - x_{s''}^{pre\ start} \geq (y_{s's''} - 1) \cdot M, s', s'' \in S, s' \neq s'', \quad (6)$$

$$x_{s'}^{pre\ start} - x_{s''}^{post\ end} + 0.5 \leq z_{s's''} \cdot M, s', s'' \in S, s' \neq s'', \text{ and} \quad (7)$$

$$x_{s'}^{pre\ start} - x_{s''}^{post\ end} \geq (z_{s's''} - 1) \cdot M, s', s'' \in S, s' \neq s'', \quad (8)$$

where M is a large positive constant.

Constraint set (5) forces $y_{s's''} = 1$ if $x_{s'}^{pre\ start} \geq x_{s''}^{pre\ start}$; the 0.5 term on the left hand side makes sure that $y_{s's''} = 1$ in case $x_{s'}^{pre\ start} = x_{s''}^{pre\ start}$. However, constraint set (5) does not force $y_{s's''} = 0$ if $x_{s'}^{pre\ start} < x_{s''}^{pre\ start}$, which is why constraint set (6) is needed. The constraint sets (7) and (8) work in the same way as (5) and (6), but for the z variables.

The constraint set

$$x_{s'}^{surg\ start} \geq x_{s''}^{post\ end} + (1 - y_{s''s'}) \cdot t^{prep} - (1 - y_{s's''}) \cdot M, s', s'' \in S_o, o \in O, s' \neq s'' \quad (9)$$

models that there should be enough time to prepare the operating rooms between each pair of subsequent surgical cases that is scheduled in the same operating room. For two surgical cases $s', s'' \in S_o$ assigned to the same operating room $o \in O$, this constraint is redundant if the surgical procedure of s'' starts later than the surgical post-procedure end time of s'; otherwise, there should be at least t^{prep} time units between these activities. For two surgical cases in sequence, we allow the surgical pre-procedure of the second surgical case to overlap with the activities of the first surgical case.

Finally, the constraints

$$\sum_{s \in S, s \neq s'} y_{s's} - z_{s's} \leq r - 1, s' \in S \quad (10)$$

make sure that no more than $r - 1$ other surgical cases are ongoing at the surgical pre-procedure start time for each of the surgical cases $s' \in S$. As described above, a surgical case s'' is ongoing at the start time of s' iff $y_{s's''} - z_{s's''} = 1$, and the number of other ongoing surgical cases at the start time of s' is therefore given by $\sum_{s \in S, s \neq s'} y_{s's} - z_{s's}$. Since s' consumes one resource, there may be at most $r-1$ other ongoing surgical cases at the start time of s'.

Computational Experiment

In a computational experiment, we apply the presented optimization model in order to generate a 1-day surgery schedule for a surgery department. In particular, we study the effect of using overlaps given a limited number of resources as described above. The purpose of the experiment is to verify the correctness of our optimization model, and to demonstrate the potential of using overlapping surgical activities. In the experiment, which is based on synthetic data, we consider three operating rooms (each room is used by one surgeon) and a given number of surgical teams. Each of the teams is allowed to assist in all of the operating rooms. In addition, it is possible to schedule one team for the surgical pre-procedure of a surgical case, whereas another team may continue to assist with the subsequent activities for that case, i.e., surgical procedure and surgical post-procedure.

For each of the operating rooms, we iteratively assign randomly generated surgical cases until the total duration of the assigned cases is at least 510 min (8.5 h), i.e., the opening hours for 1 day. For example, if we have assigned surgical cases with a total duration of 400 min, and if the next assigned surgical case has a total duration time of 110 min or more, we are done. If the next assigned surgical case instead has a total duration time less than 110 min, we generate at least one more surgical case for the considered room.

In our optimization model, a surgical case is specified using three, as afore mentioned, parameters: surgical pre-procedure time, surgical procedure time, and surgical post-procedure time. For each of the surgical cases, we randomly generate the surgical pre-procedure time as either 30 or 60 min, the surgical procedure time as either 15, 30, 45, 60, 75, 90, 105, or 120 min, and the surgical post-procedure time as either 30 or 45 min. It should be noted that we do not consider any correlation between the three parameters for a surgical case, i.e., each of the three parameters for a surgical case is generated without considering the other parameters. In Table 2, we present the surgical cases included in our experiment using the notation introduced in section "Optimization Model".

Table 2 Surgical cases included in our scenarios

Surgical case	Room	t_s^{pre}	t_s^{surg}	t_s^{post}
S1	1	60	60	45
S2	1	60	60	45
S3	1	60	60	30
S4	1	60	45	45
S5	2	60	120	45
S6	2	30	60	45
S7	2	30	115	45
S8	3	60	90	30
S9	3	30	90	30
S10	3	30	30	45
S11	3	60	115	45

In the experiment, we create (and compare) the optimal schedules for the following scenarios:

Scenario 1 (S1, baseline): No overlaps allowed. One (and only one) surgical team available for each of the operating rooms (in total, three surgical teams).
Scenario 2a (S2a): Overlaps allowed. Three surgical teams.
Scenario 2b (S2b): Overlaps allowed. Four surgical teams.

To be able to compare the generated schedules, we use the same surgical cases in all the scenarios.

For S1 (the baseline scenario), where no overlaps are allowed, we generate the optimal schedule without using the optimization model. This schedule results in additional overtime for each of the rooms: 120 (630 in total), 40 (550 in total) and 145 (655 in total) min, respectively.

For S2a and S2b, we apply our optimization model in order to generate schedules that include the possibility to use overlaps. When modeling three surgical teams (S2a), we obtain a schedule with overtimes of 75, 0, and 185 min, respectively, for the three rooms. This gives a reduction of 45 min overtime in total, compared to the situation where no overlaps are allowed (S1), even though the overtime in room three is 40 min higher in S2a. The reason for the latter is that the optimization model strives for a global optimal solution, where the aim is to minimize the total overtime in the system. When instead modeling four surgical teams (S2b), we obtain a significantly improved schedule; the overtimes in this scenario are 0, 0, and 40 min, respectively, for the three operating rooms. When comparing with S2a, the schedule in S2b gives an overtime reduction of 220 min. The generated schedules for all of the considered scenarios are provided in Fig. 5.

For modeling and solving the optimization problems, we used the AMPL modeling tool and ILOG CPLEX 10.0 solver. Scenario 2a (with three surgical teams) take approximately 26.3 h to solve, and scenario 2b (with four surgical teams) take approximately 3.7 s using a computer with 2GB memory and an Intel Pentium D CPU 2.80GHz (dual core[1]) processor with 2MB cache memory.

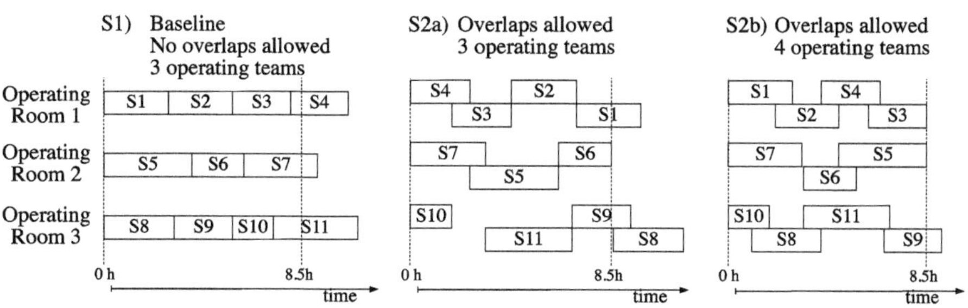

Fig. 5 Generated schedules for the three scenarios: (S1) No overlaps allowed, three surgical teams, (S2a) overlaps allowed, three surgical teams, and (S2b) overlaps allowed, four surgical teams

[1] It should be noted that our CPLEX license only allow us to use one of the cores.

Concluding Remarks

We have presented an optimization model for sequence dependent operating room scheduling in a surgery department. The optimization model differs from previous surgery scheduling models in that it includes the possibility to let the surgical pre-procedure of a surgical case overlap with the surgical procedure and surgical post-procedure activities of another, ongoing surgical case assigned to the same operating room.

In order to verify the correctness of the optimization model, we conducted a computational experiment, where we created 1-day schedules for a surgery department with three operating rooms.

The results of the experiment also indicate that allowing for overlapping surgical activities may lead to significantly improved surgery schedules, in terms of reduced overtime, compared to when no overlaps are allowed.

As discussed above, we based our experiment on synthetic data. Since the main focus of the current paper is to present our model, verify its correctness, and illustrate its potential, we find this to be sufficient. In order to make stronger conclusions about the applicability of the studied idea (of allowing overlapping surgical cases), we aim to conduct an experiment using real surgical data as part of future work.

Up to now, our focus has been on verifying the correctness of our model and establishing the potential of the idea of systematically using overlapping surgical activities. We have, therefore, not focused on the time complexity of our approach. However, the time required to find an optimal solution, emphasizes the high complexity of the considered problem. Developing customized solutions, finding methods and objective functions, which utilize on the specific characteristics of the studied problem, belongs to future work.

The presented optimization model requires that surgical cases are manually assigned to a specific day and to a specific operating room, since this is how most surgery departments operate today. The presented model will be evaluated by the surgery department at Danderyd Hospital (a university hospital in Sweden). Continuous development of the model is planned in cooperation with the hospital. However, we believe that a model that includes a mechanism for allocating surgeries to specific operating rooms and that models activities for several days may enable even better schedules to be generated. As part of future work, we aim to propose and investigate such a model.

Acknowledgements The authors would like to thank Maria Unbeck, Pamela Mazzocato, Helena Hvitfeldt-Forsberg, Andre Stark, Olof Sköldenberg and Paula KellyPettersson in the Q-hip research group from Department of Clincal Sciences, Danderyd Hospital and Medical Management Centre both Karolinska Institutet for valuable advices during the project.

References

Abdelrasol, Z.Y., Harraz, N., Eltawil, A.: A proposed solution framework for the operating room scheduling problems. WCECS 2013, 23–25 October, 2013, San Francisco, USA (2013)

Arnaout, J-P., Sevag Kulbashian, S.: Maximizing the utilization of operating rooms with stochastic times using simulation. In: Proceedings of the 2008 Winter Simulation Conference, pp. 1617–1623 (2008)

Azari-Rad, S., Yontef, A.L., Aleman, D.M., Urbach, D.R.: Reducing elective general surgery cancellations at a Canadian hospital. Can. J. Surg. **56**, 113–118 (2013)

Cardoen, B., Demeulemeester, E., Beliën, J.: Operating room planning and scheduling: a literature review. Eur. J. Oper. Res. **201**, 921–932 (2010)

Friedman, D.M., Sokal, S.M., Chang, Y., Berger, D.L.: Increasing operating room efficiency through parallel processing. Ann. Surg. **243**, 10–14 (2006)

Gupta, D., Denton, B.: Appointment scheduling in health care: challenges and opportunities. IIE Trans. **40**, 800–819 (2008)

Hans, E., Wullnik, G., van Houdenhoven, Kazemier, G.: Robust surgery loading. Eur. J. Oper. Res. **185**, 1038–1050 (2008)

Pham, D.H., Klinkert, A.: Surgical case scheduling as a generalized job shop scheduling problem. Eur. J. Oper. Res. **185**, 1011–1025 (2006)

Testi, A., Tanfani, E, Torre, G.: A three-phase approach for operating theatre schedules. Health Care Manag. **10**, 163–172 (2007)

A Mean-Field Analysis for the Two-Tiered Healthcare Network Through Nonlinear Markov Processes

Na Li, Quan-Lin Li, and Rui-Na Fan

Introduction

A healthcare delivery system in china always has a hierarchy structure, while such a whole system is generally classified into three levels: Level-I (or primary) hospitals, level-II (or secondary) hospitals and level-III (or tertiary) hospitals. Note that the level-I hospitals mainly mean community hospitals which support basic medical services for community residents; the level-II hospitals are district hospitals which provide health services for several communities; and the level-III hospitals give special medical services nationwide to address critical diseases. Furthermore, there are 6684 primary hospitals, 6722 secondary hospitals and 1851 tertiary hospitals in china by the end of June, 2014. As we know, the healthcare resource of china is very limited, since the 22 % of the world's population in china is only supported by the 3 % of the world's medical resources. For this, it is an essentially interesting problem to manage the multi-tiered hospitals whose purpose is to coordinate and to optimize the healthcare delivery service to various people.

The coordination of hospitals in china is mainly reflected by the two way referral policies. In general, most of the basic health care service is provided in the lower-level hospitals while the urgent or special health care service is given in the higher-level hospitals. In a lower-level hospital, if a patients' disease cannot be treated well, he/she has to be referred to a higher-level hospital, here we define it as

---

The original version of this chapter was revised. An erratum to this chapter can be found at DOI 10.1007/978-3-319-35132-2_20

N. Li
Department of Industrial Engineering and Management, Shanghai Jiaotong University, Shanghai, China

Q.-L. Li (✉) • R.-N. Fan
School of Economics and Management Sciences, Yanshan University, Qinhuangdao, China
e-mail: liquanlin@mail.tsinghua.edu.cn

A. Matta et al. (eds.), *Health Care Systems Engineering for Scientists and Practitioners*, Springer Proceedings in Mathematics & Statistics 169,
DOI 10.1007/978-3-319-35132-2_6

an upward referral. On the other hand, if a patient in a higher-level hospital passed their serious phase, and he/she may be referred to the lower-level hospitals for some further healing treatment, we define this process as a downward referral. Note that the upward and downward referrals motivate us in this paper to study a two-tiered healthcare network through developing a mean-field theory. We demonstrate how the upward and downward referrals highly affect performance measures of the healthcare system, which will be helpful and useful for the resource allocation of hospitals, the referral regulation and the management of patient flows.

Although the queueing theory is extensively applied to the study of healthcare systems (Lakshmi et al. 2013; Gupta 2013; Bhattacharjee and Ray 2014; Fomundam and Herrmann 2007), the traditional queueing theory is difficult to deal with the two-tiered healthcare network which is a complex and large-scale stochastic network. In this case, the paper proposes a mean-filed methodology which is established by two key techniques: a time-inhomogeneous queue and a nonlinear birth-death process. For the mean-filed theory, readers may refer to Sznitman 1989; Graham 2000; Benaim and Le Boudec 2008; Gast and Gaujal 2011; Li et al. 2014, 2015; Li and Lui 2016; Li 2014.

This paper is organized as follows. "Model Description" provides a model description for the two-tiered healthcare network. "Mean-Field Analysis" applies the mean-field theory to derive two classes of mean-filed equations. "Performance Analysis" considers a nonlinear Markov process which is established by a nonlinear birth-death process and a nonlinear QBD process. We also give performance analysis of the two-tiered healthcare network. Some concluding remarks are given in "Conclusion."

Model Description

In this section, we provide a model description for the two-tiered healthcare network, where some system parameters are given.

Now, we describe a healthcare system with upward and downward referrals, which is also a large scale queuing network, in terms of Fig. 1.

Note that the healthcare system is abstracted to two layers (in fact, the two layers of hospitals are distributed in, such as, a city or a province), the upper layer contains the higher-level hospitals such as level III hospitals, while the lower layer has the lower-level hospitals such as level II (or/and level I) hospitals.

We assume that there are N lower-level hospitals and $N\alpha$ higher-level hospitals in the whole system, where α is a positive constant between 0 and 1, which is a portion between the numbers of the higher-level hospitals and of the lower-level hospitals. For example, α is currently about 0.28 in China.

For the lower-level hospitals, the patients with less serious diseases at lower-level hospital as a Poisson process with arrival rate $N\lambda_1$. We assume that some of the less serious patients leave the system (cured or dead) with probability p_1 after an exponential service time with service rate μ_1. The others of the less serious patients go to the higher-level hospitals with probability $1 - p_1$ due to the deterioration of their disease.

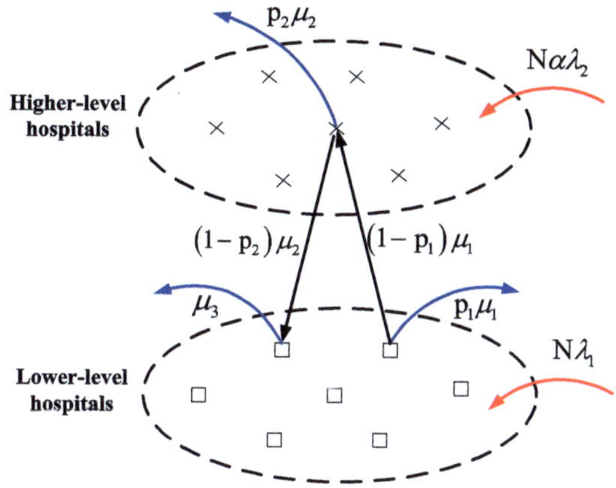

Fig. 1 Two-way referral between the higher-level hospitals and the lower-level hospitals

For the higher-level hospitals, besides the deteriorative patients' arriving, the patients with serious diseases arrive at higher-level hospitals as a Poisson process with arrival rate $N\alpha\lambda_2$. After treated by an exponential service time with service rate μ_2, the patients in the higher-level hospitals are also divided into two groups: Group-I patients are healed or dead and they leave the system; group-II patients need further recovery treatment and they are referral to lower-level hospitals. The probability that patients directly leave the higher-level hospitals is p_2. Further, we assume that the recovery patients will leave the lower-level hospitals in which the treating time is exponential distribution with service rate μ_3.

Mean-Field Analysis

In this section, we apply the mean-field theory to derive two classes of mean-filed equations, which are a key for our next study.

The Mean-Filed Scale Equations

In the lower-level hospitals, let $y_{n_1,n_2}^{(N)}(t)$ be the proportion of the lower-level hospitals with n_1 first visit patients and n_2 revisit patients in service or waiting for service. Similarly, for the higher-level hospitals, we denote by $z_k^{(N)}(t)$ the proportion of the higher-level hospitals with k patients in service or waiting for service.

For each of the higher-level hospitals, the patients' arrival rate $\overline{\lambda}_2$ can be given by two parts: One part is from the serious patients who visit directly, while another part is from the referral of the lower-level hospitals. Note that only when the lower-level hospitals are not empty, the referral patients can flow into the higher-level hospitals. Besides, since the numbers of the higher-level hospitals are less than that

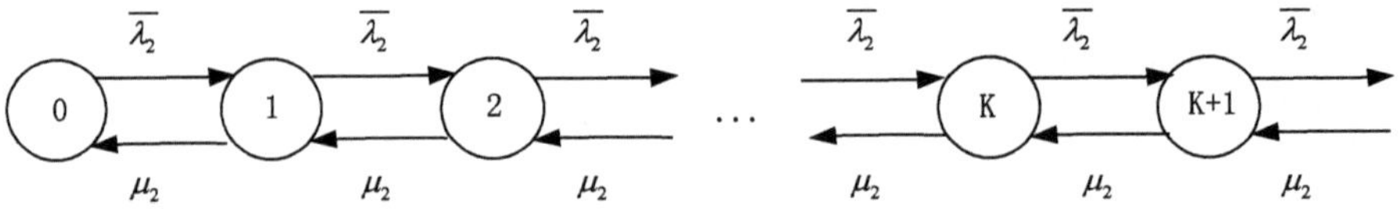

Fig. 2 The state transition relation for the queueing process in each of the higher-level hospitals

of the lower-level hospitals, the $1/\alpha$ patients of the lower-level hospitals will visit the higher-level hospitals. Therefore, using the mean-field theory, we have

$$\overline{\lambda}_2 = \lambda_2 + \frac{(1-p_1)\,\mu_1 \left(1 - \sum_{l=0}^{\infty} y_{0,l}^{(N)}(t)\right)}{\alpha}. \tag{1}$$

The state transition relation for the queueing process in each of the higher-level hospitals can be described as Fig. 2.

Based on Fig. 2, we obtain the differential equations as follows:

$$\frac{d}{dt} z_0^{(N)}(t) = -\overline{\lambda}_2 z_0^{(N)}(t) + \mu_2 z_1^{(N)}(t) \tag{2}$$

and

$$\frac{d}{dt} z_k^{(N)}(t) = \overline{\lambda}_2 z_{k-1}^{(N)}(t) - \left[\overline{\lambda}_2 + \mu_2\right] z_k^{(N)}(t) + \mu_2 z_{k+1}^{(N)}(t), \quad k \geq 1. \tag{3}$$

For each of the lower-level hospitals, there are two types of arrival: One is the direct visit patients outside the system, while the other is those recovery patients from the higher-level hospitals. In this case, the arrival rate of the recover patients is given by

$$\overline{\lambda}_1 = (1-p_2)\,\mu_2 \left(1 - z_0^{(N)}(t)\right)\alpha. \tag{4}$$

Note that when the high-level hospitals are not empty, the referred patients can flow into the lower-level hospitals. Therefore, the state transition relation for the queueing process in each of the lower-level hospitals can be described as Fig. 3.

Notice that the state transition diagram is shown in Fig. 3 where the states are arranged into levels.

For example, level 0 indicates all the transition situation about the number of less serious patients come to lower-level hospital assumed that the number of patients transfer from high-level hospital is 0.

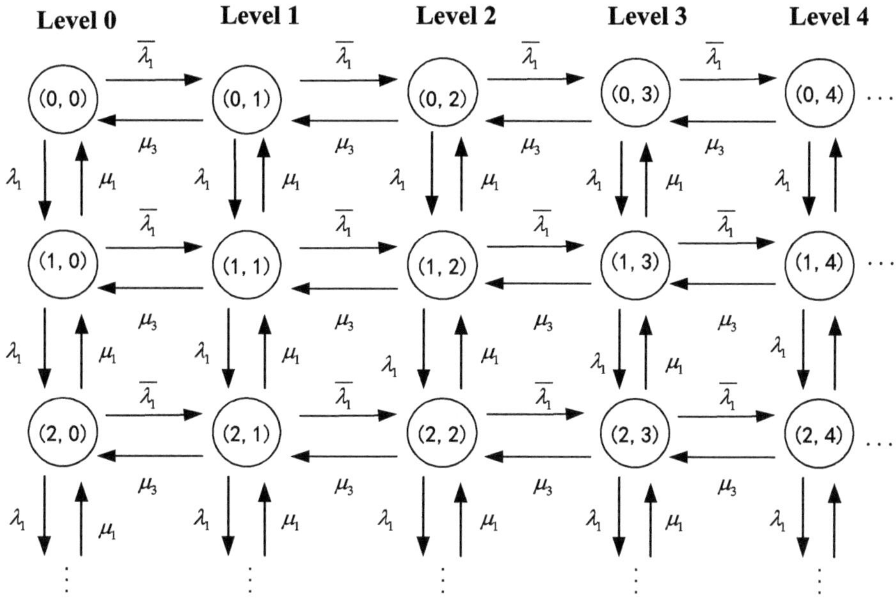

Fig. 3 The state transition relation for the queueing process in each of the lower-level hospitals

Based on Fig. 3, we obtain the differential equations as follows:
When $n_1 = n_2 = 0$,

$$\frac{d}{dt}y_{0,0}^{(N)}(t) = \mu_3 y_{0,1}^{(N)}(t) + \mu_1 y_{1,0}^{(N)}(t) - \left(\lambda_1 + \overline{\lambda}_1\right) y_{0,0}^{(N)}(t), \tag{5}$$

when $n_1 = 0, \quad n_2 = k \geq 1$,

$$\frac{\mathrm{d}}{\mathrm{d}t}y_{0,k}^{(N)}(t) = \overline{\lambda}_1 y_{0,k-1}^{(N)}(t) + \mu_3 y_{0,k+1}^{(N)}(t) + \mu_1 y_{1,k}^{(N)}(t) - \left(\mu_3 + \lambda_1 + \overline{\lambda}_1\right) y_{0,k}^{(N)}(t), \tag{6}$$

when $n_1 = k \geq 1, \quad n_2 = 0$,

$$\frac{\mathrm{d}}{\mathrm{d}t}y_{k,0}^{(N)}(t) = \lambda_1 y_{k-1,0}^{(N)}(t) + \mu_3 y_{k,1}^{(N)}(t) + \mu_1 y_{k+1,0}^{(N)}(t) - \left(\mu_1 + \lambda_1 + \overline{\lambda}_1\right) y_{k,0}^{(N)}(t), \tag{7}$$

when $n_1 = k \geq 1, \; n_2 = l \geq 1$,

$$\begin{aligned}\frac{\mathrm{d}}{\mathrm{d}t}y_{k,l}^{(N)}(t) = {} & \lambda_1 y_{k-1,l}^{(N)}(t) + \mu_3 y_{k,l+1}^{(N)}(t) + \overline{\lambda}_1 y_{k,l-1}^{(N)}(t) \\ & + \mu_1 y_{k+1,l}^{(N)}(t) - \left(\mu_1 + \mu_3 + \lambda_1 + \overline{\lambda}_1\right) y_{k,l}^{(N)}(t).\end{aligned} \tag{8}$$

For the whole system, the initial condition is given by

$$z_k^{(N)}(0) = g_k, \quad k \geq 0, \tag{9}$$

$$y_{k,l}^{(N)}(0) = h_{k,l}, \quad k, l \geq 0, \tag{10}$$

$$h_k = (h_{k,0}, h_{k,1}, h_{k,2}, \ldots), \quad h = (h_0, h_1, h_2, \ldots),$$

and let

$$(\mathrm{g}, h) = (\mathrm{g}_0, \mathrm{g}_1, \mathrm{g}_2, \ldots; h_0, h_1, h_2, \ldots)$$

be a probability vector.

The Mean-Filed Block-Structured Equations

For convenience of description, we write the matrix form of the above mean field equations (5) to (8). To that end, we write

$$\mathbf{z}^{(N)}(t) = \left(z_0^{(N)}(t), z_1^{(N)}(t), z_2^{(N)}(t), \ldots\right),$$

$$y_k^{(N)}(t) = \left(y_{k,0}^{(N)}(t), y_{k,1}^{(N)}(t), y_{k,2}^{(N)}(t), \ldots\right),$$

$$\mathbf{y}^{(N)}(t) = \left(y_0^{(N)}(t), y_1^{(N)}(t), y_2^{(N)}(t), \ldots\right).$$

For each of the lower-level hospitals,

$$\frac{d}{dt}\mathbf{z}^{(N)}(t) = \mathbf{z}^{(N)}(t)\mathbf{W}_{\mathbf{z}^{(N)}(t)}, \quad \mathbf{z}^{(N)}(0) = g, \tag{11}$$

and for each of the lower-level hospitals,

$$\frac{\mathrm{d}}{\mathrm{d}t}\mathbf{y}^{(N)}(t) = \mathbf{y}^{(N)}(t)\mathrm{V}_{\mathbf{y}^{(N)}(t)}, \quad \mathbf{y}^{(N)}(0) = h. \tag{12}$$

thus we have

$$\frac{\mathrm{d}}{\mathrm{d}t}\left(\mathbf{z}^{(N)}(t), \mathbf{y}^{(N)}(t)\right) = \left(\mathbf{z}^{(N)}(t), \mathbf{y}^{(N)}(t)\right)\begin{pmatrix} \mathrm{W}_{\mathbf{z}^{(N)}(t)} & \\ & \mathbf{V}_{\mathbf{y}^{(N)}(t)} \end{pmatrix}, \tag{13}$$

with the initial condition

$$\left(\mathbf{z}^{(N)}(0), \mathbf{y}^{(N)}(0)\right) = (g, h), \tag{14}$$

where

$$\mathbf{W}_{\mathbf{z}^{(N)}(t)} = \begin{pmatrix} -\overline{\lambda}_2 & \overline{\lambda}_2 & & & \\ \mu_2 & -\left(\mu_2 + \overline{\lambda}_2\right) & \overline{\lambda}_2 & & \\ & \mu_2 & -\left(\mu_2 + \overline{\lambda}_2\right) & \overline{\lambda}_2 & \\ & & \ddots & \ddots & \ddots \end{pmatrix},$$

$$\mathbf{V}_{\mathbf{y}^{(N)}(t)} = \begin{pmatrix} A_{0,0} & A_1 & & \\ A_{-1} & A_0 & A_1 & \\ & A_{-1} & A_0 & A_1 \\ & & \ddots & \ddots & \ddots \end{pmatrix},$$

$$A_{0,0} = \begin{pmatrix} -\left(\lambda_1 + \overline{\lambda}_1\right) & \lambda_1 & & & \\ \mu_1 & -\left(\lambda_1 + \overline{\lambda}_1 + \mu_1\right) & \lambda_1 & & \\ & \mu_1 & -\left(\lambda_1 + \overline{\lambda}_1 + \mu_1\right) & \lambda_1 & \\ & & \ddots & \ddots & \ddots \end{pmatrix},$$

$$A_1 = \begin{pmatrix} \overline{\lambda}_1 & & & \\ & \overline{\lambda}_1 & & \\ & & \overline{\lambda}_1 & \\ & & & \ddots \end{pmatrix}, \quad A_{-1} = \begin{pmatrix} \mu_3 & & & \\ & \mu_3 & & \\ & & \mu_3 & \\ & & & \ddots \end{pmatrix},$$

$$A_0 = \begin{pmatrix} -\left(\lambda_1 + \overline{\lambda}_1 + \mu_3\right) & \lambda_1 & & & \\ \mu_1 & -\left(\lambda_1 + \overline{\lambda}_1 + \mu_1 + \mu_3\right) & \lambda_1 & & \\ & \mu_1 & -\left(\lambda_1 + \overline{\lambda}_1 + \mu_1 + \mu_3\right) & \lambda_1 & \\ & & \ddots & \ddots & \ddots \end{pmatrix}.$$

Note that since $\overline{\lambda}_1$ and $\overline{\lambda}_2$ is function of $\mathbf{y}^{(N)}(t)$ and $\mathbf{z}^{(N)}(t)$, the equations (11) and (12) are a nonlinear system of differential equations.

Performance Analysis

In this section, we consider a nonlinear Markov process which is established by a nonlinear birth-death process and a nonlinear QBD process. Also, we give performance analysis of the two-tiered healthcare network.

Limiting Distribution of the Nonlinear Markov Process

For each of the higher-level hospitals, let

$$\gamma_2 = \lim_{t\to+\infty}\lim_{N\to\infty} \overline{\lambda}_2 = \lambda_2 + \frac{1}{\alpha}(1-p_1)\,\mu_1\left(1-\sum_{l=0}^{\infty}\pi_{0,l}\right).$$

Then

$$\mathbf{W}_\phi = \begin{pmatrix} -\gamma_2 & \gamma_2 & & & \\ \mu_2 & -(\mu_2+\gamma_2) & \gamma_2 & & \\ & \mu_2 & -(\mu_2+\gamma_2) & \gamma_2 & \\ & & \ddots & \ddots & \ddots \end{pmatrix}.$$

We write

$$\phi_k = \lim_{t\to+\infty}\lim_{N\to\infty} z_k^{(N)}(t).$$

Note that

$$\lim_{t\to+\infty}\left[\frac{d}{dt}z_k^{(N)}(t)\right] = 0, \quad \lim_{N\to\infty}\left\{\lim_{t\to+\infty}\left[\frac{d}{dt}z_k^{(N)}(t)\right]\right\} = 0,$$

it follows from (2) and (3) that

$$-\gamma_2\phi_0 + \mu_2\phi_1 = 0 \tag{15}$$

$$\gamma_2\phi_{k-1} - (\gamma_2+\mu_2)\,\phi_k + \mu_2\phi_{k+1} = 0, \ \ k \geq 1. \tag{16}$$

Therefore, we obtain

$$\phi_k = \left(\frac{\gamma_2}{\mu_2}\right)^k \phi_0, \quad k \geq 1,$$

where ϕ_0 is an underdetermined constant.

For each of the lower-level hospitals, we set

$$\pi_{k,l} = \lim_{t\to+\infty}\lim_{N\to\infty} y_{k,l}^{(N)}(t),\ k,l \geq 0,$$
$$\boldsymbol{\pi} = \lim_{t\to+\infty}\lim_{N\to\infty} \mathbf{y}^{(N)}(t).$$

Let

$$\mathbf{y}(t) = \lim_{N\to\infty} \mathbf{y}^{(N)}(t)$$

and

$$\mathbf{V}_{\mathbf{y}(t)} = \lim_{N\to\infty} \mathbf{V}_{\mathbf{y}^{(N)}(t)}.$$

Then it follows from (11) and (12) that

$$\frac{\mathrm{d}}{\mathrm{d}t}\mathbf{z}(t) = \mathbf{z}(t)\mathbf{W}_{\mathbf{z}(t)},\quad \mathbf{z}(0) = g, \tag{17}$$

$$\frac{\mathrm{d}}{\mathrm{d}t}\mathbf{y}(t) = \mathbf{y}(t)\mathbf{V}_{\mathbf{y}(t)},\quad \mathbf{y}(0) = h. \tag{18}$$

A similar analysis to the fixed point in Li (2015) and Li and Lui (2016), we have

$$\boldsymbol{\pi} = \lim_{t\to+\infty} \mathbf{y}(t),\quad \lim_{t\to+\infty}\left[\frac{\mathrm{d}}{\mathrm{d}t}\mathbf{y}(t)\right] = 0,$$

hence it follows from (12) and (13) that

$$\begin{aligned}\boldsymbol{\pi}\mathbf{V}_{\boldsymbol{\pi}} &= 0,\\ \boldsymbol{\pi}e &= 1 - \sum_{k=0}^{\infty}\phi_k = 1 - \tfrac{\mu_2}{\mu_2-\gamma_2}\phi_0.\end{aligned} \tag{19}$$

Since $\boldsymbol{\pi}\mathbf{V}_{\boldsymbol{\pi}} = 0$ is nonlinear, it is different from the stationary probability vector of a Markov process Q, for example, $\mathbf{x}Q = 0,\ \mathbf{x}e = 1$.

Let

$$\gamma_1 = \lim_{t\to+\infty}\lim_{N\to\infty} \overline{\lambda}_1 = (1-p_2)\,\mu_2\,(1-\phi_0)\,\alpha.$$

Then

$$\mathbf{V}_{\boldsymbol{\pi}} = \begin{pmatrix} B_{0,0} & B_1 & & & \\ B_{-1} & B_0 & B_1 & & \\ & B_{-1} & B_0 & B_1 & \\ & & \ddots & \ddots & \ddots \end{pmatrix}, \tag{20}$$

where

$$B_{0,0}=\begin{pmatrix}-(\lambda_1+\gamma_1) & \lambda_1 & & & \\ \mu_1 & -(\lambda_1+\gamma_1+\mu_1) & \lambda_1 & & \\ & \mu_1 & -(\lambda_1+\gamma_1+\mu_1) & \lambda_1 & \\ & & \ddots & \ddots & \ddots\end{pmatrix},$$

$$B_1=\begin{pmatrix}\gamma_1 & & & \\ & \gamma_1 & & \\ & & \gamma_1 & \\ & & & \ddots\end{pmatrix},\quad B_{-1}=\begin{pmatrix}\mu_3 & & & \\ & \mu_3 & & \\ & & \mu_3 & \\ & & & \ddots\end{pmatrix},$$

$$B_0=\begin{pmatrix}-(\lambda_1+\gamma_1+\mu_3) & \lambda_1 & & & \\ \mu_1 & -(\lambda_1+\gamma_1+\mu_1+\mu_3) & \lambda_1 & & \\ & \mu_1 & -(\lambda_1+\gamma_1+\mu_1+\mu_3) & \lambda_1 & \\ & & \ddots & \ddots & \ddots\end{pmatrix},$$

Corresponding to (19), we set

$$\begin{aligned}\boldsymbol{\pi} &= (\boldsymbol{\pi}_0, \boldsymbol{\pi}_1, \boldsymbol{\pi}_2, \ldots),\\ \boldsymbol{\pi}_0 &= (\pi_{0,0}, \pi_{0,1}, \pi_{0,2}, \ldots),\\ \boldsymbol{\pi}_1 &= (\pi_{1,0}, \pi_{1,1}, \pi_{1,2}, \ldots),\\ \boldsymbol{\pi}_2 &= (\pi_{2,0}, \pi_{2,1}, \pi_{2,2}, \ldots),\\ &\vdots\end{aligned}$$

Then

$$\boldsymbol{\pi}_k = \boldsymbol{\pi}_1 R^{k-1}, \quad k \geq 1,$$

where $\boldsymbol{\pi}_1$ and $\boldsymbol{\pi}_2$ are uniquely given by

$$\begin{cases}\boldsymbol{\pi}_0 B_{0,0} + \boldsymbol{\pi}_1 B_1 = 0,\\ \boldsymbol{\pi}_0 B_{-1} + \boldsymbol{\pi}_1 (B_0 + RB_{-1}) = 0,\\ \boldsymbol{\pi}_0 e + \boldsymbol{\pi}_1 e = 1.\end{cases} \tag{21}$$

Since

$$\sum_{k=0}^{\infty} \phi_k + \sum_{k=0}^{\infty} \pi_k e = 1,$$

we have

$$\phi_0 + \phi_1 + \phi_2 + \cdots = 1 - \boldsymbol{\pi}_0 e - \boldsymbol{\pi}_1 (I-R)^{-1} e,$$

which follows

$$\phi_0 = \left(1 - \frac{\gamma_2}{\mu_2}\right)\left[1 - \boldsymbol{\pi}_0 e - \boldsymbol{\pi}_1 (I - R)^{-1} e\right].$$

Performance Computation

In each of the higher-level hospitals, the stationary mean of patients is given by

$$\begin{aligned} N^{(H)} &= \sum_{k=1}^{\infty} k\phi_k = \sum_{k=1}^{\infty} k\left(\frac{\gamma_2}{\mu_2}\right)^k \phi_0 = \frac{\mu_2\gamma_2}{(\mu_2-\gamma_2)^2}\phi_0 \\ &= \frac{\gamma_2}{\mu_2-\gamma_2}\left[1 - \boldsymbol{\pi}_0 e - \boldsymbol{\pi}_1 (I - R)^{-1} e\right]. \end{aligned}$$

In each of the lower-level hospitals, the stationary mean of patients of type one is given by

$$N_1^{(L)} = \sum_{k=1}^{\infty} k \sum_{l=0}^{\infty} \pi_{k,l} = \sum_{k=1}^{\infty} k\boldsymbol{\pi}_k e = \boldsymbol{\pi}_1 \sum_{k=1}^{\infty} kR^{k-1} e = \boldsymbol{\pi}_1 \frac{1}{(1-R)^2} e,$$

and the stationary mean of patients of type two is given by

$$N_2^{(L)} = \sum_{l=1}^{\infty} l \sum_{k=0}^{\infty} \pi_{k,l}.$$

Conclusion

This paper provides a two-tiered healthcare network for analyzing and designing multi-tiered healthcare systems used in many cities of China. Note that the two-tiered healthcare network is a complex and large-scale system; hence its analysis is always difficult and challenging. In the situation, this paper proposes a mean-field theory for studying the two-tiered healthcare network, where the mean-field theory is developed by two key techniques: a time-inhomogeneous queue and a nonlinear birth-death process. Based on this, we give some highlight on the healthcare service delivery system management, and the mean-field methodology of this paper can be applied to dealing with more general healthcare systems in practice, such as, multi-tiered healthcare networks, heterogeneous healthcare systems, and role of different geography in healthcare systems.

Acknowledgement This work was supported by the National Natural Science Foundation of China under grant No.71471114, No.71432006, No. 71271187, No. 71471160, and the Fostering Plan of Innovation Team and Leading Talent in Hebei Universities under grant No. LJRC027.

References

Benaim, M., Le Boudec, J.Y.: A class of mean-field interaction models for computer and communication systems. Perform. Evaluat. **65**, 823–838 (2008)

Bhattacharjee, P., Ray, P.K.: Patient flow modelling and performance analysis of healthcare delivery processes in hospitals: a review and reflections. Comput. Ind. Eng. **78**, 299–312 (2014)

Fomundam, S., Herrmann, J.: A survey of queuing theory applications in healthcare. Working Paper. Digital Repository at the University of Maryland (2007)

Gast, N., Gaujal, B.: A mean field approach for optimization in discrete time. Discrete Event Dyn. Syst. **21**(1), 63–101 (2011)

Graham, C.: Chaoticity on path space for a queueing network with selection of the shortest queue among several. J. Appl. Probab. **37**, 198–201 (2000)

Gupta, D.: Queueing models for healthcare operations (Chapter 2). In: Denton, B.T. (ed.) Handbook of Healthcare Operations Management: Methods and Applications. International Series in Operations Research & Management Science, vol. 184, pp. 19–44. Springer, New York (2013)

Lakshmi, C., Sivakumar, A.I.: Application of queueing theory in health care: a literature review. Oper. Res. Health Care **2**, 25–39 (2013)

Li, Q.L., Lui, J.C.S.: Block-structured supermarket models. Discrete Event Dyn. Syst. **26**(2), 147–182 (2016)

Li, Q.L.: Tail probabilities in queueing processes. Asia Pac. J. Oper. Res. **31**(2), 1–31 (2014)

Li, Q.L., Dai, G., Lui, J.C.S., Wang, Y.: The mean-field computation in a supermarket model with server multiple vacations. Discrete Event Dyn. Syst. **24**(4), 473–522 (2014)

Li, Q.L., Du, Y., Dai, G., Wang, M.: On a doubly dynamically controlled supermarket model with impatient customers. Comput. Oper. Res. **55**, 76–87 (2015)

Sznitman, A.: Topics in propagation of chaos. In: Springer Lecture Notes in Mathematics 1464, École d'Été de Probabilités de Saint-Flour XI, pp. 165–251(1989)

Scheduling Magnetic Resonance Imaging Examinations: An Empirical Analysis

Filippo Visintin and Paola Cappanera

Introduction

In the last 30 years, many of the Organisation for Economic Co-operation and Development (OECD) members countries have experienced an ever increasing growth in the demand for diagnostic imaging (DI) services, such as Computed Tomography (CT) and Magnetic Resonance Imaging (MRI) examinations (OECD 2011). Unfortunately, there is a widespread mismatch between supply and demand that, in the long run, causes diverging waiting lists, long waiting times and a generalised patient dissatisfaction. In an attempt to control these criticalities policy makers have set specific targets in terms of Maximum Time Before Treatment (MTBT) (see, for example, Canadian Association of Radiologists (2013)). At the same time, the interest of practitioners and scholars on this challenging topic is witnessed by quite a high number of recent contributions in the scientific literature on medical appointment scheduling models (Cayirli and Veral 2003; Li and Benton 2003; Gupta and Denton 2008; Carpenter et al. 2011).

The case study this work focuses on, concerns the scheduling of MRI examinations for outpatients in a leading Italian Hospital that for confidentiality reasons will be referred to as Alpha. However, the problem treated is quite general and the issues considered are common to several medical appointment scheduling problems. In this study, patients on the hospital waiting list are characterised by the following

F. Visintin (✉)
IBIS LAB, Dipartimento di Ingegneria Industriale, University of Florence,
Viale Morgagni 44, 50134 Florence, Italy
e-mail: filippo.visintin@unifi.it

P. Cappanera
IBIS LAB, Dipartimento di Ingegneria dell'Informazione, University of Florence,
Via Santa Marta 3, 50134 Florence, Italy

A. Matta et al. (eds.), *Health Care Systems Engineering for Scientists and Practitioners*, Springer Proceedings in Mathematics & Statistics 169,
DOI 10.1007/978-3-319-35132-2_7

five attributes: *priority class*, *type*, *duration*, *sedation*, and *contrast*. The *priority class* attribute corresponds to the Mean Time Before Treatment (MTBT) expressed in days (*class A* 10 days, *class B* 30 days and *class C* 60 days). The *type* attribute, instead, identifies the type of examination required (*neuro* type if the examination concerns the nervous system, *body* type otherwise); correspondingly radiologists are characterized by a speciality (neuro or body) which relates to the type of examinations they are skilled to perform. *Duration* indicates the number of time slots required by an examination. Finally, *sedation* and *contrast*, are 0-1 attributes that indicate, respectively, if a patient needs to be sedated and if the examination requires the use of a contrast agent. In case of sedation the test will require the presence of an *anaesthetist*.

Alpha adopts a *dynamic* First Come First Served (FCFS)-like approach to allocate patients to the two scans, one of which has come into operation recently: specifically, patients are assigned the first available appointment at the time they call. However, contrary to a pure FCFS approach, a certain amount of capacity (*reserved capacity*) is allotted to class A patients to manage urgencies.

This study consists in an empirical analysis, which has twofold aim: to investigate the long-term effect of the implementation of a FCFS-like heuristic; to assess the impact on the capability of matching the patient's due date (1) cross training radiologists across examination types, (2) reserving capacity to urgent patients, (3) reducing the number of inappropriate prescriptions.

To this aim, we developed a simulation model, which generates the demand for MRI examinations and schedules each patients upon arrival using a greedy algorithm. Such an algorithm finds, for each incoming patient, the first available slots, i.e., the first series of consecutive time-slots where all the resources needed to perform the test are concurrently available. Service times are considered deterministic and expressed in time slots. The patients' arrival process is considered as a Homogeneous Poisson Process. Patients mix and the arrival rate were estimated by the hospital management. The analysis considers a time horizon of 1 year.

The simulation model monitors the evolution of the waiting times and of other relevant indicators such as number of patients scheduled, the number of patient scheduled on late, the delay, etc., on a longitudinal basis.

The major findings of this study can be summarised as follows:

1. the effects on waiting times of a given appointment scheduling policy must be observed in the long run
2. reserving capacity to urgent patients allows matching due-dates for these patients but at the expenses of the other ones
3. cross-training radiologists, helps matching the due-dates of all patient types
4. reducing demand by avoiding inappropriate prescriptions produces a significant reduction in the number of the patients scheduled on late.

The remainder of the paper is organised as follows: in Sect. 2, we provide a brief review of the literature. In Sect. 3, we present the characteristics of the addressed problem. In Sect. 4, we describe the greedy heuristic adopted and the simulation model, while in Sect. 5, we present the experimental campaign we have carried out and we discuss the empirical results. Finally, conclusions are drawn in Sect. 6.

Literature Review

Contributions on MRI topic are spread over a number of scientific disciplines; here we briefly review some of the recent studies from operations research and management science areas. Specifically, we only focus on those contributions that according with the classification criterion proposed in Gupta and Denton (2008), consider *indirect waiting time* as a performance indicator. Indirect waiting time is the time elapsed between the patient's call for an appointment and the scheduled appointment time and it is opposed to *direct waiting time* which measures the time spent by the patient in the clinic on the day of the appointment.

Open access and advance scheduling policies, which offer every patient a same-day appointment or a scheduled future appointment at a convenient time, have been recently investigated to control and reduce long waiting times. In this context, the following problems have been addressed: the optimal sizing of the patient panel via queueing models (Green and Savin 2008), the patients' choice behaviour via a Markov decision process (Gupta and Wang 2008), the *dynamic* appointment scheduling problem via heuristic methods based on a Markov decision process (Liu et al. 2010). Here *dynamic* means that the appointment schedule is determined upon the system state and it is used in contrast to *static* where all the decisions are made a priori, thus disregarding the current state of the system.

Among the static approaches, Granja et al. (2014) combine a simulated annealing heuristic with a discrete event simulation to address the elective patient admission scheduling problem with the objective of minimizing patients' waiting time and maximizing throughput. A comparison between an open access scheduling policy and a fixed appointment policy is investigated in Qu and Shi (2009): here different strategies are proposed to determine how to allocate the capacity both to the short-notice appointments and to appointments fixed ahead of time.

Problem Addressed

As pointed out in the introduction, we assume that each patient is characterised by five attributes *priority class*, *type*, *duration*, *sedation*, and *contrast*. Each patient needs to be scheduled within a certain *due date*. Scheduling a given patient requires the simultaneous availability of several resources, namely: (1) a scan for the entire duration of the examination; (2) a radiologist mastering the skill required by the type of examination (body/neuro); (3) an anaesthesiologist and a nurse for the entire duration of the examination if a patient requires sedation; (4) a nurse for the entire duration of the examination if a patient requires the use of a contrast agent. For instance, a patient needing a body MRI with contrast will require a body radiologist and a nurse; If the same patient needs also to be sedated than an anaesthesiologist will be required as.

The arrival times of the MRI requests is assumed to follow a Homogeneous Poisson Process (HPP) distribution with intensity equal to λ. Information about λ as well as about the probability of occurrence of each possible combination of the five mentioned attributes, were provided by the hospital management. In addition, the hospital management stated that, by sensitizing doctors, there is the possibility to reduce the number of inappropriate prescriptions. We assume that the hospital effort to reduce inappropriate prescriptions results in a lower value of λ. We consider two scans and no backlog, i.e. no patients in the waiting list at the beginning of the simulation. We assume to schedule incoming patients using the heuristic algorithm described in the next section that reproduces the appointment scheduling practices used at Alpha.

Simulation Model and Implemented Heuristics

The simulation model was realized in Visual Basic for Application and implemented in Excel. For each simulation run, the model generates all the requests that will arrive in the incoming year reproducing a HPP of intensity λ. Each arriving request is assigned with an array of attributes according to the probabilities provided by Alpha. The generated waiting list is saved in a database. The model browses the requests in chronological order and schedule them upon resources availability. More specifically, if the request regards a class A patient, the algorithm checks for the availability of a set of consecutive time slots where all the resources needed to perform the examination are contemporarily available. The search is done at first within the slots reserved to class A patients, and then, if it fails, it is extended to all of the other slots. For class B and class C patients, instead, the search for suitable slots is done only within not reserved slots. Regardless of her/his priority class, when a patient is scheduled her/his request' status switches from *pending* to *fulfilled*, and the status of the resources operating in the identified slots switch from *available* to *unavailable*. In case of cross-training, the availability of radiologists is assessed considering all the radiologists on-duty independently from their specialty.

Empirical Results

In our experimental campaign, we considered a time horizon of 51 weeks and we explored the effect of three experimental factors, namely Cross Training (CT), Reserved Capacity (RC) and Prescription Appropriateness (PA), each characterised by two levels. Factors and factor levels are described in Table 1.

For each experimental point, we performed 20 simulation runs lasting 51 weeks (which implied running each algorithm $51 \times 20 = 1020$ times). All the runs assume no backlog. For each scheduled patient we computed the waiting time, i.e. the number of days between the day when the patient calls the hospital to obtain an

Table 1 Experimental design

Factor	Level "0"	Level "1"
Cross training (CT)	Radiologists own a specialty only (either NEURO or BODY)	Radiologists can perform both NEURO and BODY tests
Reserved capacity (RC)	No capacity is reserved to class A patients	One session per week is reserved to class A patients
Prescription appropriateness (PA)	Arrival times follow a HPP process with $\lambda = 18.0$	Arrival times follow a HPP process with $\lambda = 17.5$

appointment and the day when the examination is actually scheduled. Based on the waiting time, it is possible to check whether the patient is processed on time or on late and, in the latter case, to compute the delay, i.e. the positive difference between the number of days a patient waited and the number of day s/he was expected to wait, according to her/his priority class. The results of this study are presented as follows.

First, we investigate how the aforementioned performances evolve over time, focussing on the base scenario where all factors are at their low level. Second, we investigate the (main and interaction) effects of each factor, on two keys system performance, namely:

1. the total number of patients scheduled in the planning horizon;
2. the total number of patients scheduled on late in the planning horizon.

Given the massive amount of experimental data, the results are presented in graphical form.

Base Scenario (CT = 0, RC = 0, PA = 0), Longitudinal Analysis

Figure 1 reports the boxplot of the number of patients scheduled for each week of the planning horizon stratified by priority class. As can be noticed, such a number increases the first 2 weeks and then stabilizes. This behaviour is due to the fact that at the beginning of the time horizon, there is no patient in queue and thus the number of patients to schedule is actually smaller than the capacity. After a transient of about 2 weeks, the mean number of patients scheduled stabilizes at around 110 patients per week.

While the number of patients scheduled stabilizes after the first 2 weeks, the time waited before being examined increases monotonically in the time horizon, for all the priority classes, which implies that the steady state is not reached in the considered time-horizon. Figures 2 and 3 display respectively the boxplot of waiting time and of the delay stratified by priority class.

Looking at Figs. 2 and 3 it clearly emerges that for the first 9 weeks almost all patients are scheduled on time. Then the system accumulates some delay for Class A patients and starting from week 34, also for the class B ones. Class C patients,

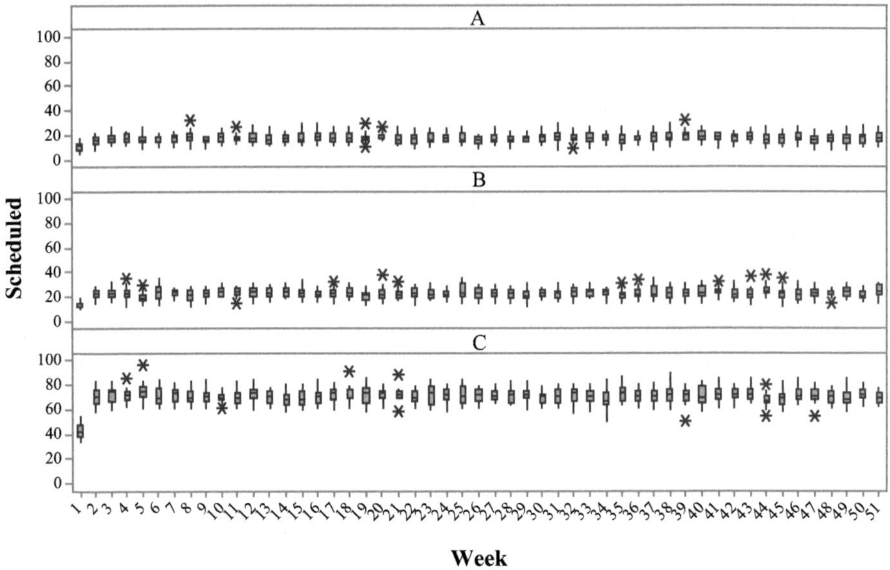

Fig. 1 Boxplot of number of patients scheduled

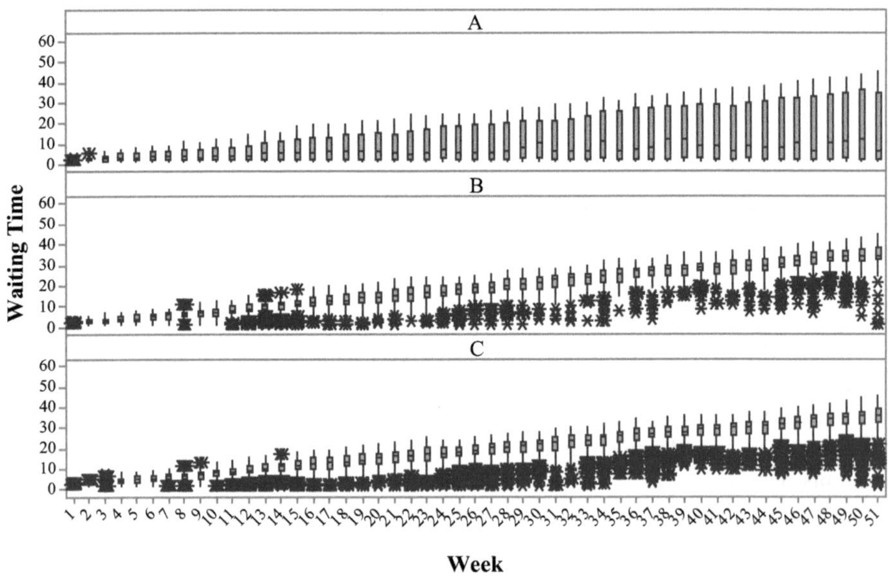

Fig. 2 Boxplot of waiting time

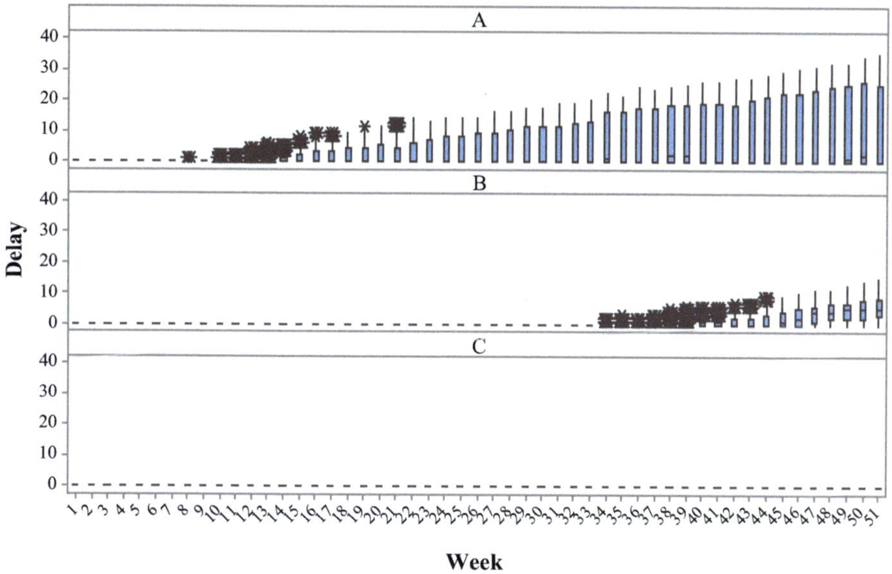

Fig. 3 Boxplot of delay

instead, experience no delay. For class A patients, despite the median value of the delay is always close to zero, delays can assume extreme values that are definitely not acceptable (the third quartile of delay at week 51 is bigger than 25 days).

In addition to measure the extent of the delay, it is of interest also to calculate the percentage of patients that experience a delay of at least 1 day. Such a percentage is hereafter referred to as % On Late. Figure 4 displays the boxplot of % On Late stratified by priority class. As can be noticed not only class A patients experience excessively long delays, but also the percentage of patients that is processed on late is way too large (the third quartile of % On Late is around 50 % starting from week 17).

Factorial Analysis

To better understand the main and interaction effects of each factor, we performed four 2^3 experiments obtained using two different output variables and two different sets of patients. In particular, the output variables are: (i) the total number of patients scheduled in the planning horizon, and (ii) the total number of patients scheduled on late in the planning horizon. The sets of patients considered instead are (i) the class A patients only and (ii) all the patients aggregated.

For all the fitted general linear models, we checked the assumptions of homoscedasticity with the Levene's test and normality of the residuals with

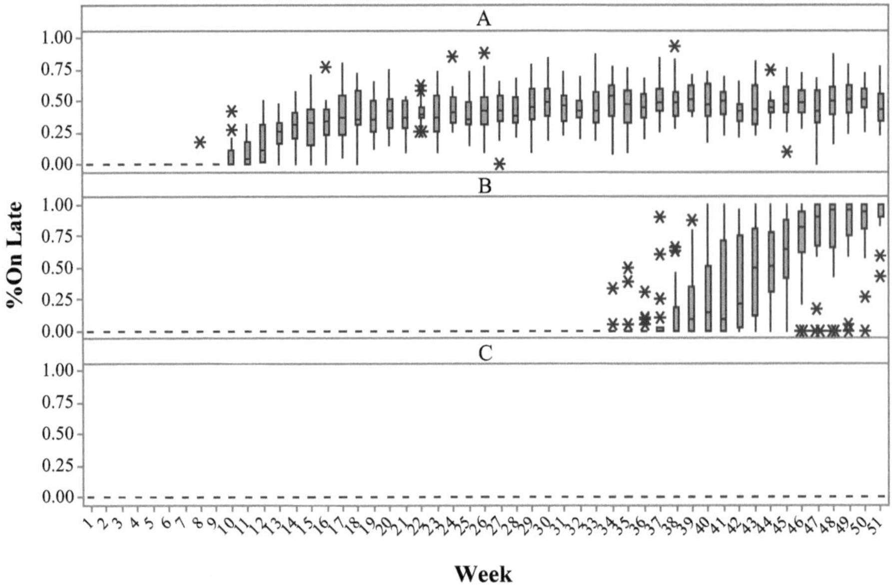

Fig. 4 Boxplot of % of patients scheduled on-late

the Anderson Darling's test. For all the tests we were unable to reject the null hypotheses ($p > 0.05$). Due to space constraints, we will not report the results of all the tests in full, instead, we will show some relevant main effect and interaction plots.

Figure 5 refers to the class A patients, and plots the mean effects of RC, CT and PA on the number of patients scheduled. Figure 6, instead, displays the number of class A patients scheduled on late. Each dot of these graphs represents the average value across scenarios assumed by the output variable when the factors are at their low level (0) and at their high level (1). As can be noticed, for RC and CT, a shift from level 0 to level 1, determines an increase in the number of class A patients scheduled while for PA, as obvious, it determines a decrease. In addition, for all factors moving from level 0 to 1 determines a dramatic reduction in the number of patients scheduled on late. Looking at Fig. 7, which refers to the number of class A patients scheduled on late, however, we can observe a significant and negative interaction between RC and CT. Such an interaction implies that the extent to which reserving capacity leads to a reduction of the number of class A patients scheduled on late depends on whether radiologists are cross-trained or not. In particular, when radiologists are cross-trained, reserving capacity to class A patients has a very limited impact.

While for class A patients reserving capacity leads to a performance improvement, if we consider all patient types aggregated, we can see that reserving capacity to class A patients is, indeed, counterproductive. In fact, looking at Fig. 8 we can

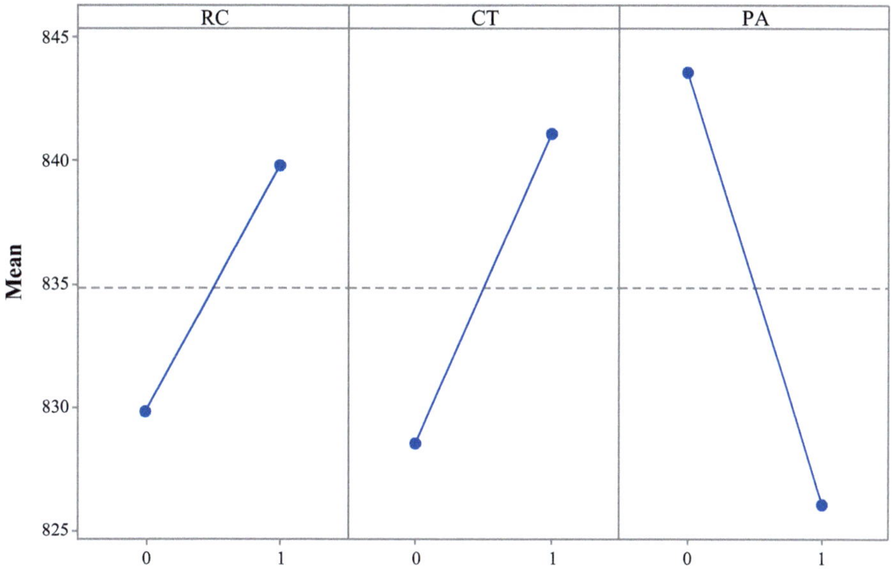

Fig. 5 Main effects plots of RC, CT and PA on class A patients scheduled

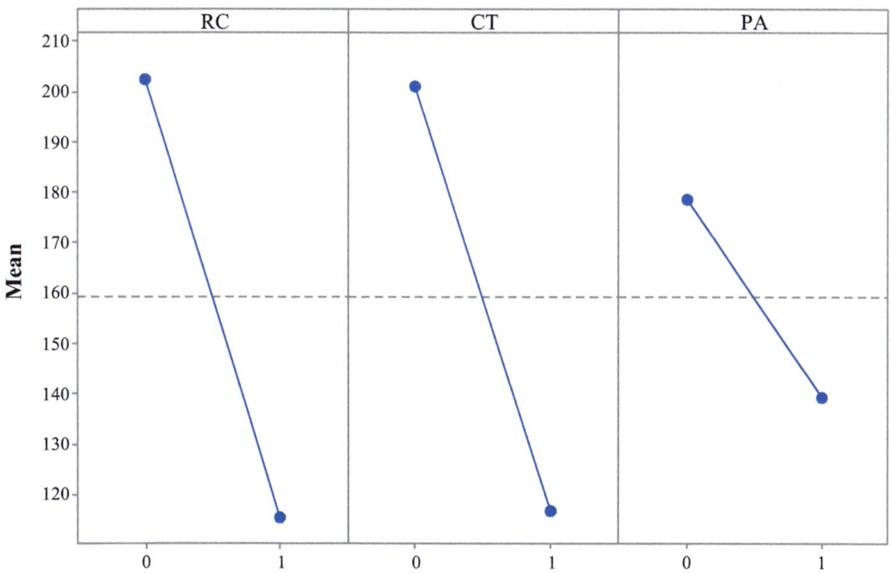

Fig. 6 Main effects plots of RC, CT and PA on class A patients scheduled on late

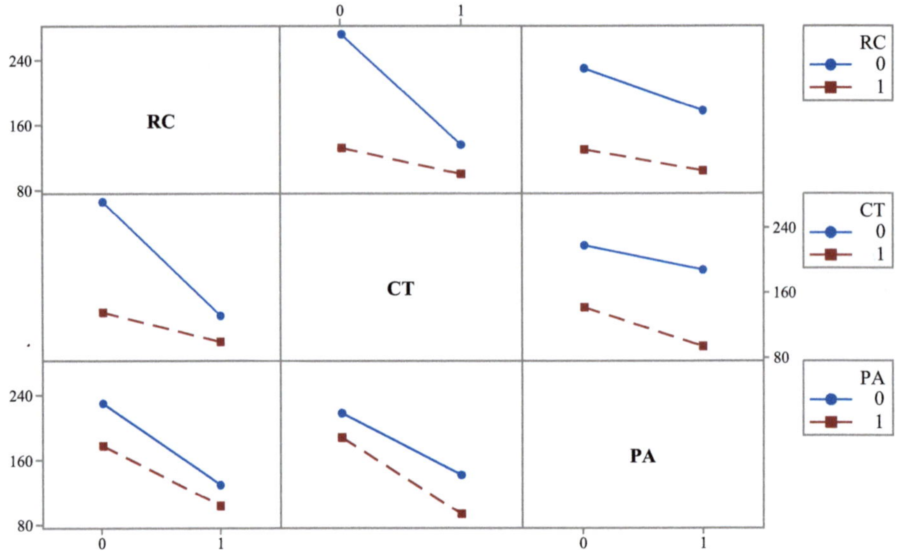

Fig. 7 Interaction effects of RC, CT and PA on class A patients scheduled on late

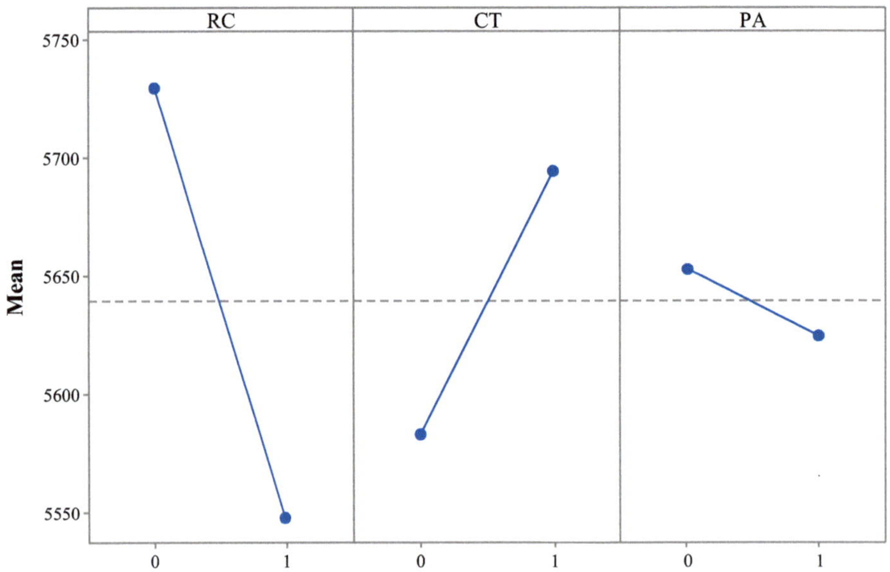

Fig. 8 Main effects plots of RC, CT and PA on patients scheduled

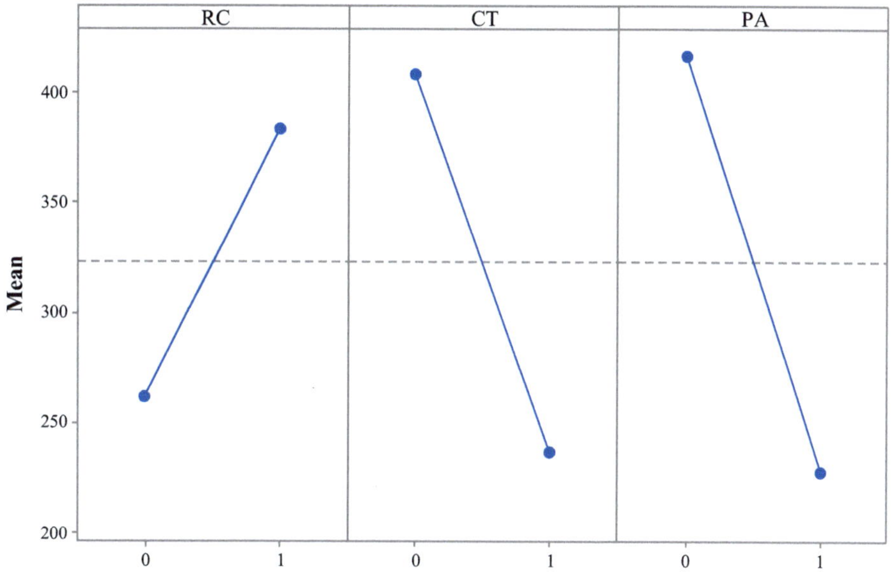

Fig. 9 Main effects plots of RC, CT and PA on patients scheduled on late

notice a negative main effect of RC on the number of patients scheduled (i.e. on average reserving capacity leads to less patients scheduled) and a positive main effect of RC on the number of patients scheduled on late (i.e. on average, reserving capacity leads to more patients scheduled on late). Finally, from Fig. 10 we can recognize a significant interaction between RC and CT and between RC and PA. These interactions imply that the negative impact of RC on the number of patients scheduled on late is more relevant when CT and PA are, respectively, at their high and low level.

It also worth to observe the magnitude of the main effect of PA (see Figs. 6 and 9) on the patient scheduled on late. Even a moderate decrease of λ (in our case, from 18 to 17.5 patients/day), in fact, has a dramatic impact on the number of patients scheduled on late especially when we reserve capacity to urgent cases.

Conclusion

The preliminary results presented in this paper provide some insight on how to evaluate and improve the efficiency of a MRI service. First, this study clearly demonstrates that the effects on waiting times of a given appointment scheduling policy must be observed in the long run. Most of the study available in the literature propose empirical analyses that are limited to very short time horizon. In this study if we had considered an 8-week time horizon we would have concluded that the system

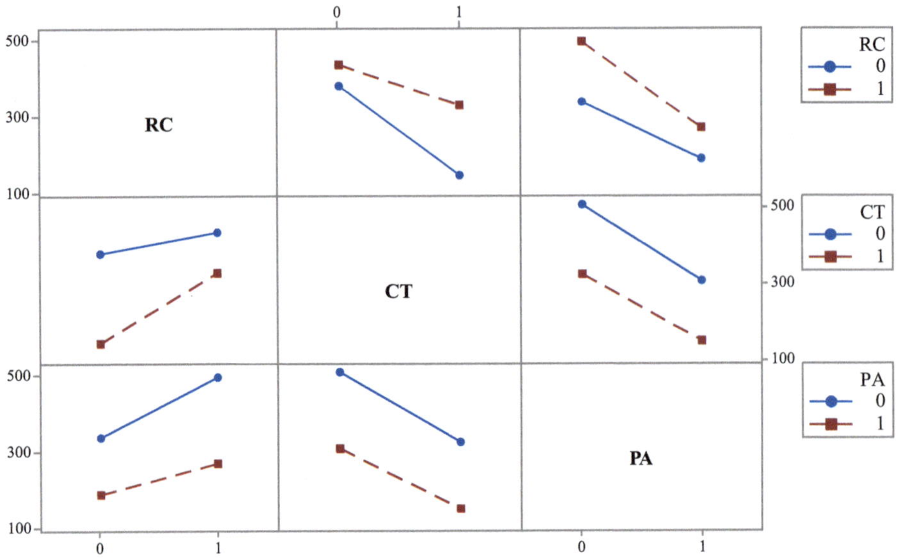

Fig. 10 Interaction effects of RC, CT and PA on patients scheduled on late

is capable of matching the patients due date, but it is not true. Second, reserving capacity to urgent patients allows matching due-dates for these patients but, as obvious, at the expenses of the other ones. Cross-training radiologists, instead, helps matching the due-dates of all patient types, by increasing the throughput. As such, if it is not too costly to implement, such a practice should be preferred. Third, the arrival intensity has a strong effect on all performances. The system is very sensitive to variations of λ. As such, reducing demand e.g. by avoiding inappropriate prescriptions produces a significant reduction in the number of patients scheduled on late.

This study is not without limitations. First, it investigates one hospital setting only (the Alpha one), which clearly limits the study external validity. Second, referring to the Alpha setting, for all the investigated scenarios, we never reach a steady state (see Figs. 2 and 3). Since the time horizon is rather long (1 year) it probably means that, even with the most favourable (yet reasonable) combination of factor levels the intensity of the arrivals is not compatible with the system capacity. Our future research effort will thus devoted to: (1) extending the number of investigated settings; (2) testing the effectiveness of an offline scheduling approach. This latter approach assumes that when patients call for an appointment, they are not immediately assigned with an appointment. Instead, they are provided with an expected waiting time. The MRI examination scheduling is then performed offline using and optimization model. Subsequently, patients are called back and given an appointment, according to the model solution. By doing so, we expect to increase the patients' throughput and to prevent waiting times to diverge.

References

Canadian Association of Radiologists. National maximum wait time access targets for medical imaging (2013)

Carpenter, A.P., Leemis, L.M., Papir, A.S., Phillips, D.J., Phillips, G.S.: Managing magnetic resonance imaging machines: support tools for scheduling and planning. Health Care Manag. Sci. **14**(2), 158–173 (2011)

Cayirli, T., Veral, E.: Outpatient scheduling in health care: a review of literature. Prod. Oper. Manag. **12**(4), 519–549 (2003)

Granja, C., Almada-Lobo, B., Janela, F., Seabra, J., Mendes, A.: An optimization based on simulation approach to the patient admission scheduling problem using a linear programing algorithm. J. Biomed. Inform. (2014). doi:10.1016/j.jbi.2014.08.007

Green, L.V., Savin, S.: Reducing delays for medical appointments: a queueing approach. Oper. Res. **56**(6), 1526–1538 (2008)

Gupta, D., Denton, B.: Appointment scheduling in health care: challenges and opportunities. IIE Trans. **40**(9), 800–819 (2008)

Gupta, D., Wang, L.: Revenue management for a primary-care clinic in the presence of patient choice. Oper. Res. **56**(3), 576–592 (2008)

Li, L., Benton, W.: Hospital capacity management decisions: Emphasis on cost control and quality enhancement. Eur. J. Oper. Res. **146**(3), 596–614 (2003)

Liu, N., Ziya, S., Kulkarni, V.G.: Dynamic scheduling of outpatient appointments under patient no-shows and cancellations. Manuf. Serv. Oper. Manag. **12**(2), 347–364 (2010)

OECD: Organisation for Economic Co-operation and Development: Number of magnetic resonance imaging (MRI) units and computed tomography (CT) scanners, Health Data 2011 (2011)

Qu, X., Shi, J.: Effect of two-level provider capacities on the performance of open access clinics. Health Care Manag. Sci. **12**(1), 99–114 (2009)

A Managerial Use of the Volume-Outcome Association for Hospital Planning

Arianna Alfieri, Elisabetta Listorti, and Andrea Matta

Introduction

Ensuring global quality within healthcare systems is one of the biggest challenges policy makers are facing today. The presence of an uneven distribution of outcome quality levels is increasingly reported within national contexts and it raises concerns about its causes (Birkmeyer et al. 2004). Even if external determinants like patient characteristics and procedure complexity may produce differences in physicians' performances, there are several inner factors that can negatively affect the outcome as well, such as resource availabilities, treatment appropriateness, and most of all organizational aspects of the healthcare structures (Birkmeyer et al. 2004).

How to improve the outcome of healthcare organizations by reorganizing the hospitals operating in a territory is the topic of this paper. We focus on surgery, which is a dynamic specialty whose competency consists of a combination of technical skill, team performance, communication and decision making (Sarker and Vincent 2005). All these factors affect the surgical patient pathway and make research in the field of surgery extremely interesting from a managerial perspective. Moreover, to support health policy makers in planning decisions, both the meaning of quality service has to be defined and the organizational factors influencing outcomes have to be identified.

Since it is highly dependent on the receivers' perceptions of needs (Eldabi et al. 2002), within the healthcare sector there is not a univocal interpretation of what

A. Alfieri (✉) • E. Listorti
Politecnico di Torino, Torino, Italy
e-mail: arianna.alfieri@polito.it; elisabetta.listorti@polito.it

A. Matta
Shanghai Jiao Tong University, Shanghai, China
e-mail: matta@sjtu.edu.cn

A. Matta et al. (eds.), *Health Care Systems Engineering for Scientists and Practitioners*, Springer Proceedings in Mathematics & Statistics 169,
DOI 10.1007/978-3-319-35132-2_8

constitutes a quality service. However, despite this general lack of consensus, there are some areas where performances can be measured through widely recognized outcome markers. Indeed, the mortality rate index can be used as an outcome measurement of surgical operations (Donabedian 1966). Moreover, since the mortality rate is a direct outcome measurement, it has immediate validity and it is likely to get the greatest approval from surgeons, who are further subjected to the Hawthorne effect (i.e., they improve their performance in response to their awareness of being observed) (Birkmeyer et al. 2004).

As for the influencing factors, within the healthcare quality framework, an essential finding has been represented by the identification of an existing relation between volumes and outcomes (Luft et al. 1979). There is a recurring trend that associates higher volume of activity to better results, which in turn means better patients' health conditions (Birkmeyer et al. 2002; Halm et al. 2002). Better results for higher volumes are especially plausible within surgery wards: high-volume hospitals may have more surgeons who specialize in specific procedures, more consistent processes for postoperative care, better-satisfied intensive care units, and greater resources, in general, for dealing with postoperative complications (Birkmeyer et al. 2002). Examples of research conducted on the volume-outcome association are Piano Nazionale Valutazione Esiti (PNE) in Italy, created by Agenas (Amato et al. 2012), whose objective is to assess healthcare service quality levels; the Leapfrog group in US, which has undertaken the definition of minimum thresholds that guarantee an admissible quality level for health outcomes (Birkmeyer and Dimick 2004). All the studies mainly reveal that there is a positive effect of volume on outcomes for each medical procedure, although its extent varies depending on the clinical area itself. For example, less frequent clinical conditions, which usually are the most difficult, show largely better outcomes if performed volumes increase. However, as these difficult procedures are rare, the achieved total number of people who benefits from the better outcomes is not a relevant percentage of the whole population.

Health planning problems have been approached so far using mostly operational or tactical decision levels combined with pure managerial purposes and objectives (security, accessibility, productivity and so on) (Kokangul 2008; Nguyen et al. 2005; Lovejoy and Li 2002). As a consequence, more attention has been focused on monitoring waiting times, throughput, utilization, leveling, patient deferrals, etc. (Cardoen et al. 2010) rather than on clinical outcomes of healthcare organizations. Other works are related to resource dimensioning of healthcare structures. Nonetheless, throughout the literature, dimensioning problems often limit the definition of ward dimension to the 'number of beds' choice, referring to a spatial perspective. Dimensioning actually entails a clinical perspective as well since it corresponds also to the number of operations completed, which not always coincides with the number of beds.

In this paper, we take the clinical perspective as we focus on the volume of activity to be performed within a ward. Hence, our planning decision is to establish the optimal number of interventions to be allocated to different surgery wards. To determine the optimal ward dimension, we use the volume-outcome relation, which allows us to consider outcomes as well. Although improving patients' health conditions is the only aim in policy decisions and thus also in our research, which exact objective to pursue is not clear among the several stakeholders. In fact, the requirement of improving patients' health conditions can be expressed in different ways. Among the various possibilities, we consider minimizing the mortality rate and minimizing the mortality index. These objectives sound similar, nevertheless, they implicitly give a different type of priority to health conditions. Using one or the other can drive the decision maker towards a different volume allocation (of each surgery intervention) to territorial hospitals, as it will be shown in the following sections. Therefore, we examined these two objective functions to shed some light on the possible consequences of strategic decisions based on structural variables like ward and hospital dimensions.

The reminder of the paper is structured as follows. Section "Problem Description" formally introduces the problem, the main notation and the different objective functions. The analysis of the objective function behaviors is reported in section "Mortality vs. Mortality Rate". Section "Case Study" presents a case study on a provincial hospital configuration. Section "Conclusions" concludes the paper.

Problem Description

Our problem concerns the strategic decision of volume allocation of surgery interventions among hospitals. The formalization of such a problem, however, is not simple since it depends on the point of view we take. In fact, a healthcare system is composed by a variety of actors. In particular, the Italian context accounts for four main categories: commissioners, providers, physicians and patients. Commissioner is who demands healthcare services and it is usually an institutional figure (e.g., a Central Government or a Local Government). The commissioners are expected to care about population health and therefore they ask providers for healthcare services. Providers answer through the supply of the requested services. Physicians are the experts who deal with patients, who in turn receive the service. Each actor has its own interests and perspectives and therefore it is relevant to keep into consideration their different behaviors and interactions. In this paper, we have taken the commissioner perspective because during the dimensioning decisional process the commissioner is taking the first step. Hospitals dimensioning is indeed a strategic problem that does not directly involve neither physicians nor patients. Provider concerns must be taken into account. Nonetheless, as dimensioning deals with territorial healthcare configuration and people health needs, it is mainly an institutional duty to allow new structures to be built and dimensioned.

We consider a geographic region with I hospitals denoted by index $i = 1, \cdots, I$ and J surgical operations denoted by index $j = 1, \cdots, J$. The actual request of operations for surgery j for the given geographical area is denoted by d_j.

The planning problem has a stochastic nature. Dimensioning decisions are in fact based on forecasts, and even when they reproduce real population needs, patients still have the possibility either to choose in which hospital they want to have surgery or to choose not to have surgery. This possibility causes a difference between the forecasted number of operations (what is strategically planned) and the performed ones (what actually occurs) for each hospital. However, in order to focus primarily on the volume-outcome association, in this paper we deal with the deterministic case, i.e., we assume that what we plan will surely occur. Thus, variable $x_{ij} \in R^+$ indicates the volume of surgical operation j allocated and performed in hospital i.

We denote by m_{ij} the mortality rate of people receiving operation j in hospital i. The mortality rate is defined as the ratio between the number of dead patients during a given time interval and the total number of treated patients during the same time interval. It is a function of the number of treated patients (i.e., $m_{ij}(x_{ij})$). However, the functional form linking mortality rate and volume depends on the type of surgery. Finally, each surgery j has a marginal cost a_j and, at the same time, there is a fixed regional funding b available to pay for all the performed operations. Our dimensioning problem can be mathematically represented as follows:

$$max \;\; H \tag{1}$$

$$\text{s.t.} \;\; \sum_{i,j} a_j x_{ij} \leq b \tag{2}$$

$$\sum_{i} x_{ij} \leq d_j \qquad \forall j \tag{3}$$

$$l_i \leq \sum_{j} x_{ij} \leq u_i \qquad \forall i \tag{4}$$

$$\alpha_j \leq \sum_{i} x_{ij} \leq \beta_j \quad \forall j \tag{5}$$

Function (1) is the objective function and will be discussed below. Constraint (2) simply states that the total cost related to the planned volume of operations cannot exceed the available monetary resources. Constraints (3) bound the number of performed operations not to exceed the population requests. These constraints are relevant due to the alarming problem of overtreatment, which has been spreading during the last decade (American Board of Internal Medicine Foundation 2016). The volume of operations allocated to hospitals is subjected to capacity constraints (4). If we give a null value to l_i and infinite value to u_i, we allow for the extreme possibilities to open and/or close hospitals. Finally, ethic or epidemiological reasons can impose to treat a minimum number of patients, or to not exceed a maximum limit, for different types of operations, as represented by thresholds α_j and β_j in constraints (5).

The objective function (1) deals with the maximization of patients' health conditions. *Better health for all* is the objective declared by all the institutions involved in the healthcare sector. This goal can be translated into different functional forms. We considered two possibilities. The first one represents the minimization of the total mortality index:

$$\min z_1 = \sum_{i,j} x_{ij} * m_{ij}(x_{ij}) + \sum_{j}[(d_j - \sum_{i} x_{ij})^+ * \gamma_j] \quad (6)$$

This function includes two terms. The first term is the expected perioperative mortality. The second term, instead, represents the people who died because not operated. In other words, this second group of people requested to have surgery but did not receive any treatment within the considered hospitals. Parameter γ_j can have a double interpretation: it can be seen as the mortality rate of these patients who are to be redirected to other structures; or it can correspond to a penalty rate that is given to the defaulting hospitals.

The second objective function aims at reducing the average mortality rate:

$$\min z_2 = \frac{1}{IJ} \sum_{i,j} m_{ij}(x_{ij}) \quad (7)$$

This function promotes a fair attention for people living in different areas as it takes into account the average mortality rate of each kind of operation within the considered hospitals.

Even if the two objective functions seem interchangeable, they drive to different solutions. As it will be shown in section "Mortality vs. Mortality Rate", the first objective function endorses the quality of the performed operations, while the second one prefers the equity of the provided service. Referring to the goal *better health for all*, the first objective function seems to stress the adjective *better*, while the second objective function focuses on the part *for all.*

Mortality vs. Mortality Rate

In this section, we focus on the two objective functions to get insight in how their behaviors (and thus their proposed solutions) differ. We use the mortality curves taken from the Piano Nazionale Valutazione Esiti (PNE), whose database contains information collected from all the Italian hospitals on operation outcomes.

We considered the dimensioning problem, described in the previous section, in four basic cases: (1) one hospital, one operation type, (2) two hospitals, one operation type, (3) one hospital, two operation types, (4) two hospitals, two operation types.

The two operation types we used in our analysis are the Coronary Artery Bypass Graft Surgery (CABG) and R operation. Volume-outcome relation for CABG is reported in Fig. 1. As the range of volume from 0 to 60 is subjected to high variability, we considered only the decreasing part of the CABG mortality curve. Instead, R is an artificial operation created to represent an operation that has the same trend as CABG but with higher mortality rates. The reason why we artificially created R operation is to avoid misleading interplays between the two curves.

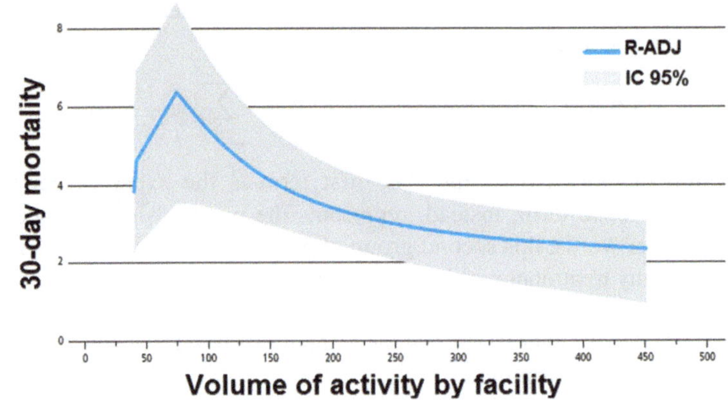

Fig. 1 Aortocoronary bypass; analysis of the association between 30-day mortality and volume of activity by facility, Italy 2011 (Amato et al. 2012)

Case 1 is trivial, since both the objective functions drive to perform as many operations as possible. This is reasonable since the more operations are performed within a single structure, the lower is the value of total mortality (z_1), the better is the mortality rate (z_2).

Case 2 considers one operation and two hospitals. In-depth analysis of the objective functions shows opposite behaviors. In fact, z_1 reaches its minimum when it leads to unbalanced solutions, allocating to one hospital a higher volume than to the other. Instead, z_2 has minimum values when it evaluates symmetrical solutions, with a balanced allocation between hospitals. This behavior is expected since z_1 aims at minimizing the total number of deaths, thus preferring to concentrate a high number of operations in one hospital to decrease the mortality rate for the higher number of patients (it does not consider if those operated in the other hospital will have a higher risk, since they will be a quite small number). On the other hand, z_2 works for an equal mortality rate between different structures, thus it fosters an equal distribution of demand. Figure 2a reports the trends assumed by z_1 and z_2 when the total demand d is equal to 300. Asymmetrical allocation to hospital 1 (either when x_{11}=60, which means $(d - x)$ is allocated to hospital 2, or vice versa when x_{11}=240) results in opposite objective functions values. In fact, unbalanced solutions drive to z_1 optimal value, while symmetrical solutions guarantee the minimum for z_2.

Case 3 considers two operations and one hospital. Again, we observe z_1 assuming lower values with an unbalanced allocation, which is emphasized by z_1 favoring the less risky operation. Oppositely, z_2 is minimized preferring a more equal distribution. Figure 2b reports z_1 and z_2 trends when $d = 300$. Asymmetrical allocation to hospital 1 is further fostered by z_1, which strongly favors the less risky intervention. On the contrary, z_2 promotes an equal allocation among operations, which is slightly unbalanced due to the higher amount of R operations that allows a lower mortality rate (thus more similar to the CABG one).

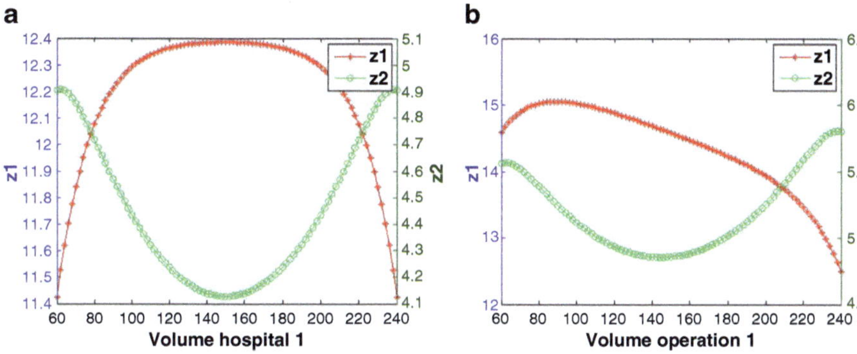

Fig. 2 Different objective functions trends, case 2 (**a**) and case 3 (**b**)

Case 4 deals with two types of operations and two hospitals. The analytical description of the objective functions is less intuitive in this case, thus we postpone it to the case study, where we consider the more general situation of multiple hospitals and multiple operations (of which case 4 is a particular case). However, as it will be clear in the next section, its features are recalling the previous cases.

Case Study

The case study considers a reconfiguration of some Piedmont Region surgery wards. We have considered five types of surgical interventions for which there is evidence of the volume-outcome association (as shown in Fig. 3): aortocoronary bypass, surgery for colon cancer, surgery for pancreatic cancer, surgery for lung cancer, nonruptured abdominal aortic aneurysm (aaa). We considered three hospitals that are present in Turin in order to analyze a provincial configuration. In order to focus specifically on the volume-outcome association, we did not take into consideration the cost constraint (2). For each hospital, we assumed a total capacity equal to the total number of the five types of operations actually performed. We have considered the possibility of closing wards and hospitals, thus we set the hospital capacity lower bounds to zero. As for α_j and β_j, we have imposed both as lower and upper bound the number of required interventions, in order to satisfy the total demand and to avoid overtreatment.

We looked for the optimal volume allocation and eventually we compared our solutions to the actual real configuration. For actual volume allocation we used the real data available from the PNE analysis of 2013 (Agenas 2014).

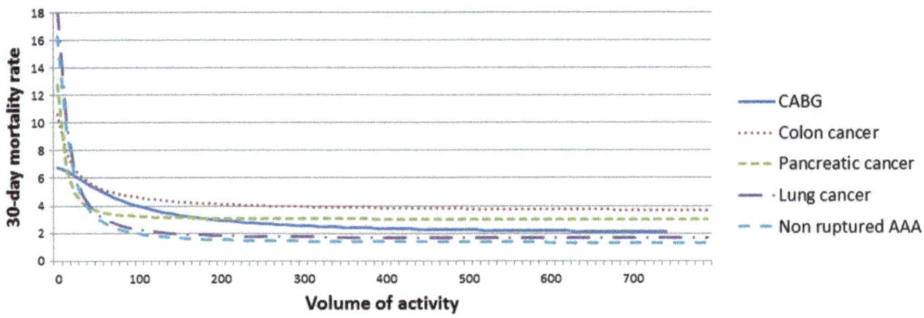

Fig. 3 Mortality curves for five main surgical operations (Amato et al. 2012)

Taking into consideration the Region as a closed system, we observed how commissioners answer to population needs. Above all, we compared different quality levels that can be provided to patients depending on the volume allocation choices.

We have tested two scenarios. The first scenario deals with an allocation made from greenfield. We do not have constraints, apart from the hospital total capacity. The solution obtained using z_1 clearly concentrates the interventions in single structures. A similar suggestion is obtained in case of z_2 minimization, as it was expected since the possibility of opening and closing wards drives toward the trivial case one hospital/one operation. Analyzing the differences between proposed and actual provincial configurations in terms of operation volumes, it clearly appears the suggested specialization of hospitals in specific interventions, rather than an apparently unjustified division of interventions among structures.

The second scenario has an additional constraint concerning the adaptability of surgeons and operating rooms, which better (although not exhaustively) represents reality. Hospitals keep on performing the types of operation they have performed the previous year. In fact, surgeons usually specialize in one kind of interventions and hospitals too as a consequence. In this scenario, z_1 still drives to specialization, especially for the riskier interventions. Instead, z_2 promotes a more balanced allocation, confirming the behavior observed in section "Mortality vs. Mortality Rate". However, both z_1 and z_2 lead to a higher total mortality (as compared with the first scenario), because of the impossibility to freely allocate operations. Hence, specialization constraints prevent from optimal solutions.

Figure 4 shows the three configurations obtained minimizing z_1 (actual, proposed from greenfield, proposed with specialization constraints) in terms of volume allocation between hospitals, while Fig. 5 reports the related mortality and mortality rates values calculated for each type of intervention.

The proposed configuration from greenfield drives to the lowest values of total mortality and mortality rates. Interestingly, the configuration with the better outcome requires several changes in the territorial organization, since hospitals need

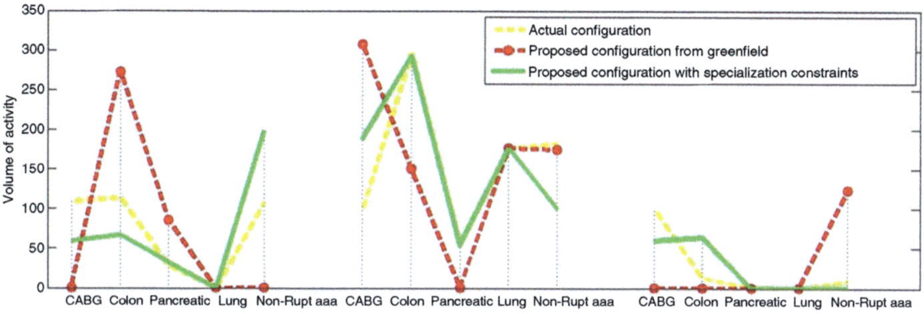

Fig. 4 Objective function z_1: volume allocation between hospitals in different configurations

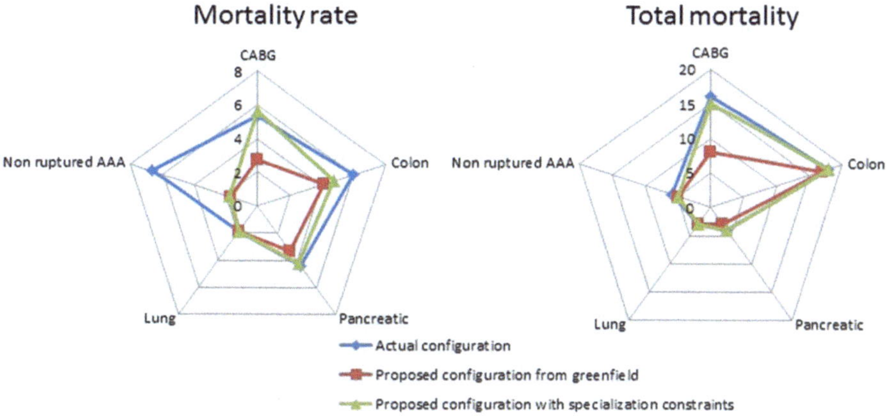

Fig. 5 Comparison between average mortality rates and total mortality of each type of intervention in the three configurations

to re-arrange their wards in order to concentrate on some operations (not necessarily performed in the hospital before the reconfiguration), neglecting others (i.e., closing the related wards). It is also noteworthy the worse outcome caused by the proposed configuration with specialization constraints.

Despite the several simplifying assumptions, the case study is of extreme interest because it shows a concrete action policy makers could take in the Piedmont Region. The model suggests to concentrate surgical interventions in specialized hospitals, where volumes and thus quality levels are high. The regional consequences are not to be ignored, both for hospitals that would host more patients and hospitals that would be forced to close. Also, concentration could lead to unfair outcomes in different areas of the territory.

Conclusions

In this paper, we have investigated the potentialities of the volume-outcome relation. The managerial use of the volume-outcome association represents an innovative starting point for planning decisions. It definitely allows policy makers to consider together managerial and medical factors in the decision process so as to obtain high quality levels in health outcomes.

We have used two functional forms to translate the general objective *better health for all*. The first translation focused on the minimization of the number of dead patients, while the second one interpreted better patient conditions as a low average mortality rate. In this way, we have deepened the implications of the use of volume-outcome curves, demonstrating the importance of dealing with care with the strategic use of clinical factors for planning processes.

Despite the interesting findings, our study is only preliminary and there are several limitations and additional factors that need to be taken into account in the future research. As for factors affecting quality in hospitals, even if volume is the most recognized one, there are studies that show how the performance can be influenced by other elements, e.g., the complexity of cases treated by hospital or the volume performed by single surgeons, and their impact should be deeply analyzed. As for the obtained configurations, we are aware of the external constraints policy makers have to face. Hospital configurations should in fact consider not only the effectiveness of treatments but also their costs, and above all the geographical distribution and the accessibility to healthcare services (Amato et al. 2012). Finally, the point of views and the interactions of the different actors have to be considered.

References

Agenas (National Agency for regional health services): National Outcome Program (2014). Available online http://95.110.213.190/PNEed14

Amato, L., et al.: Volume and health outcomes: evidence from systematic reviews and from evaluation of Italian hospital data. Epidemiol. Prev. **37**, 1–100 (2012)

American Board of Internal Medicine Foundation (2016): Choosing Wisely. Available online at http://www.choosingwisely.org/

Birkmeyer, J.D., Dimick, J.B.: Potential benefits of the new Leapfrog standards: effect of process and outcomes measures. Surgery **135**, 569–575 (2004)

Birkmeyer, J.D., Siewers, A.E., Finlayson, E.V., Stukel, T.A., Lucas, F.L., Batista, I., et al.: Hospital volume and surgical mortality in the United States. N. Engl. J Med. **346**, 1128–1137 (2002)

Birkmeyer, J.D., Dimick, J.B., Birkmeyer, N.J.O.: Measuring the quality of surgical care: structure, process, or outcomes? J. Am. Coll. Surg. **198**, 626–632 (2004)

Cardoen, B., Demeulemeester, E., Beliën, J.: Operating room planning and scheduling: a literature review. Eur. J. Oper. Res. **201**, 921–932 (2010)

Donabedian, A.: Evaluating the quality of medical care. Milbank Mem. Fund Q. **44**(3), 166–206 (1966)

Eldabi, T., Zahir, I., Ray, J.P.: A proposed approach for modeling health-care systems for understanding. J. Manag. Med. **16**, 170–187 (2002)

Halm, E.A., Lee, C., Chassin, M.R.: Is volume related to outcome in health care? A systematic review and methodologic critique of the literature. Ann. Intern. Med. **137**, 511–520 (2002)

Kokangul, A.: A combination of deterministic and stochastic approaches to optimize bed capacity in a hospital unit. Comput. Methods Prog. Biomed. **90**, 56–65 (2008)

Lovejoy, W.S., Li, Y.: Hospital operating room capacity expansion. Manag. Sci. **48**, 1369–1387 (2002)

Luft, H.S., Bunker, J.P., Enthoven, A.C.: Should operations be regionalized? the empirical relation between surgical volume and mortality. N. Engl. J. Med. **301**, 1364–1369 (1979)

Nguyen, J.M., Six, P., Antonioli, D., Glemain, P., Potel, G., Lombrail, P., Le Beux, P.: A simple method to optimize hospital beds capacity. Int. J. Med. Inform. **74**, 39–49 (2005)

Sarker, S.K., Vincent, C.: Errors in surgery. Int. J. Surg. **3**, 75–81 (2005)

A Discrete Event Simulation Model for the Admission of Patients to a Home Care Rehabilitation Service

Azadeh Maroufkhani, Ettore Lanzarone, Cecily Castelnovo, and Maria Di Mascolo

Introduction

An important rising of the Home Care (HC) sector has been observed in all of Western countries during the last years, due to demographic and epidemiological changes, and cost reduction constraints. Indeed, the demographic trend of recent years has been characterised by a significant life expectancy increase, resulting into both an increased demand and a higher care complexity due to the presence of chronic degenerative diseases and co-morbidities. This is also combined with a social fragility phenomenon, i.e., families are fragmented and care services are often provided to sick and poor patients who live alone. Institutionalisation can not be the solution, because of high costs and low quality of life for patients. In response, there are nowadays several types of HC providers, ranging from general providers to specialised ones that take care of the multidimensional needs of specific patient classes. Eligible patients can choose the provider among the available ones accredited for the specific service required.

Managing and coordinating an efficient care delivery are difficult tasks, which require suitable skills and tools. The different planning levels of a general HC provider are discussed in Lanzarone et al. (2012), and the literature presents several tools and approaches to support all of these levels. However, in providers that take care of specific patient classes, there is an additional task that is not appropriately

A. Maroufkhani • M.D. Mascolo
Université Grenoble Alpes and CNRS, G-SCOP, F-38000 Grenoble, France
e-mail: az.maroufkhani@gmail.com; Maria.Di-Mascolo@grenoble-inp.fr

E. Lanzarone (✉) • C. Castelnovo
Istituto di Matematica Applicata e Tecnologie Informatiche (IMATI), Consiglio Nazionale delle Ricerche (CNR), Milan, Italy
e-mail: ettore.lanzarone@cnr.it; cecily.castelnovo@email.it

A. Matta et al. (eds.), *Health Care Systems Engineering for Scientists and Practitioners*, Springer Proceedings in Mathematics & Statistics 169,
DOI 10.1007/978-3-319-35132-2_9

addressed, yet. This is the *admission* process, in which a new patient is evaluated and admitted to the service if he/she matches the provider characteristics, if the criteria required by the health system are all respected, and if the provider has enough free capacity to admit him/her. This task is of particular importance when patients require a fast beginning of the care pathway in order to get the desired benefits from the therapy. In this paper, we focus on the HC Rehabilitation (HCR) service. Usually, a patient is eligible for a specific set of HCR providers if he/she is in an acute condition, or for another set if in a chronic condition. The first case is a relevant example of fast care beginning, as described above.

The aim of our work is to provide a quantitative tool, based on Discrete Event Simulation (DES), for analysing and optimising the admission process in HC services. The choice of a simulation model rather than other techniques, e.g., queueing networks, is motivated by the high versatility we want for the tool, in particular in the choice of the probability density functions for the random parameters (e.g., treatment duration and patient's conditions) and in the fine modelling of the admission policies. In this paper, we present a first DES model for the admission of new patients to HCR services, and we apply it to the case of a provider operating according to the regulation of the health system of Milan, Italy (Deliberation IX/3851 of Region Lombardia 2012; Regional Decree of Lombardia 2012). This case is however general enough to derive considerations for the entire class of HCR providers.

The paper is structured as follows. A brief literature review on HC and the admission problem is provided in section 2. The description of the problem addressed in this paper is presented in section 3, and the DES model is outlined in section 4". Then, the application of the DES model to a set of instances and the results are shown in section 5. Conclusions and future extensions are finally given in section 6.

Literature Review

The literature on HC management can be divided into two groups: a first group deals with the daily scheduling of visits, personnel routing and patient assignment, whereas a second group deals with staff planning and the management in a mid- and long-term perspective. The admission problem is a complementary task within the first group. In this group, a large number of studies analyse the HC scheduling problem alone (Bertels and Fahle 2006; Eveborn et al. 2006) or, in a patient centred approach, focus on patient assignment under continuity of care (Lanzarone et al. 2012; Hertz and Lahrichi 2009; Yalçındağ et al. 2012; Lanzarone and Matta 2012, 2014; Carello and Lanzarone 2014; Yalçındağ et al. 2014). However, the admission problem together with the management of waiting lists is only marginally addressed, with qualitative approaches (Meiland et al. 1996) or by means of a linear stock-flow model (Kommer 2002). Considering the specific HCR case, several literature studies deal with the clinical benefits (Wijkstra et al. 1995; Wiles et al. 2001) or the cost-effectiveness analysis (Mitchell et al. 2005) of this service, whereas management issues are not addressed in detail. Finally, considering the methodology adopted in this paper, DES represents an useful tool largely and successfully adopted in many healthcare management problems, as documented in some reviews (Jacobson et al. 2006; Günal and Pidd 2010).

Problem Description

We address the case of HCR delivered to non-ambulatory patients who live in Milan, Italy. This case is representative of a general class of HCR services for acute patients, for which the admission of new patients is a fundamental task. The goal is to ensure a fast admission associated with the right care pathway. Two main issues arise: non-legitimate admission requests and resource saturation due to the limited number of physiotherapists. The latter leads to the presence of the waiting lists. It is highly important to properly manage waiting lists: on the one hand, the patient may have serious damages if the care intervention is delivered too late; on the other hand, a long waiting list could determine a shift towards private solutions, thus reducing equity in accessing health services.

The service is delivered to: patients who have suffered from a *recent* trauma (e.g., fall or fracture); patients who are suffering from neurodegenerative diseases; chronic patients (e.g., Parkinson's patients). Indeed, the main acceptance criteria are:

- patient unable to go out autonomously;
- architectural barriers preventing the patient to go out;
- non-transportable patient, e.g., for severe respiratory failure;
- presence of a caregiver and suitable home conditions;
- one of the following conditions, specific for rehabilitation: post-acute patient with recent trauma or recent hospitalisation period; entrapment syndrome due to a recent acute event; neurodegenerative disease (e.g., amyotrophic lateral sclerosis).

Moreover, no more than 60 days must be elapsed from the event when referred as recent, and patients with a Barthel index (Mahoney and Barthel 1965; Collin et al. 1988; Sainsbury et al. 2005) above 80 can not be taken in the charge. Both an individual rehabilitation project (PRI) and an individual rehabilitation program (pri) prepared by a physiatrist are required. If the patient already has these documents, just a general practitioner's prescription for a HCR cycle is needed to activate the service. Otherwise, a second prescription for a physiatric visit is also required to get PRI and pri.

The steps of the admission process are outlined in Fig. 1. The process starts with a phone call from the patient (or his/her caregiver) to an accredited HCR provider. The call is handled by a front office, which consists of both clinical staff, who helps the patient to find the best solution for his/her needs, and administrative staff, who

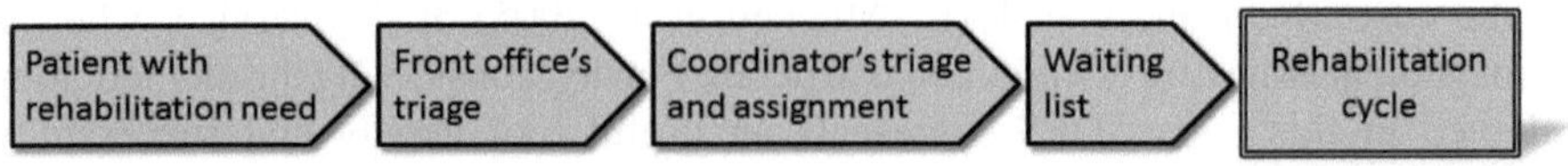

Fig. 1 Phases of the admission process in the considered HCR provider

records all personal data and verifies the presence of all mandatory documentation (PRI, pri and prescription). If acceptance criteria are respected but some documents lack, the front office gives all information necessary for their proper acquisition. Typically, in this latter case, the patient or his/her caregiver gets the documents and after a few days it is possible to proceed. If the request seems appropriate and when the documentation is complete, the call is forwarded to the HCR service coordinator.

The next phase consists of a pre-assessment made by the coordinator, i.e., a triage to verify that all clinical criteria are met. If the patient is eligible, then the coordinator assigns a physiotherapist based on workload levels and districting criteria. However, it may happen that there is no immediate availability, and the patient is inserted in a waiting list.

One of the coordinator's tasks is to ensure the shortest waiting time. He/she evaluates the waiting time and communicates the service starting date to the patient or the caregiver, who can decide whether to wait or to ask to another provider. In any case, the waiting time cannot exceed 60 days from the acute event, also including the days elapsed between the event and the first phone call to the provider (usually no more than 15 days). Finally, as soon as the assigned physiotherapist can take the patient in the charge, the coordinator calls back the patient (or the caregiver) to confirm the first visit. If the patient did not move to another provider, the physiotherapist performs a last assessment during the first visit. If neither the patient had complications nor his/her Barthel index moved above 80, the HCR cycle can start.

Discrete Event Simulation Model

In this section, we briefly present the structure and the steps of the admission process included in the DES model. The provider is modelled as front office, coordinator, and physiotherapists performing visits. The admission process, outlined in Fig. 2, starts with a phone call from the patient or his/her caregiver to the HCR provider.

The front office receives the calls and acts as a first filter. Different types of calls are received, including those for new HCR activations. It is first verified whether the patient meets all criteria and holds the general practitioner's prescription for a HCR cycle. The presence of PRI and pri is also checked; they are mandatory documents, but if they lack, it is sufficient a second prescription from the patient's general practitioner with the request for a physiatrist specialistic visit. Then, with this second prescription, PRI and pri are prepared by a physiatrist of the same HCR facility.

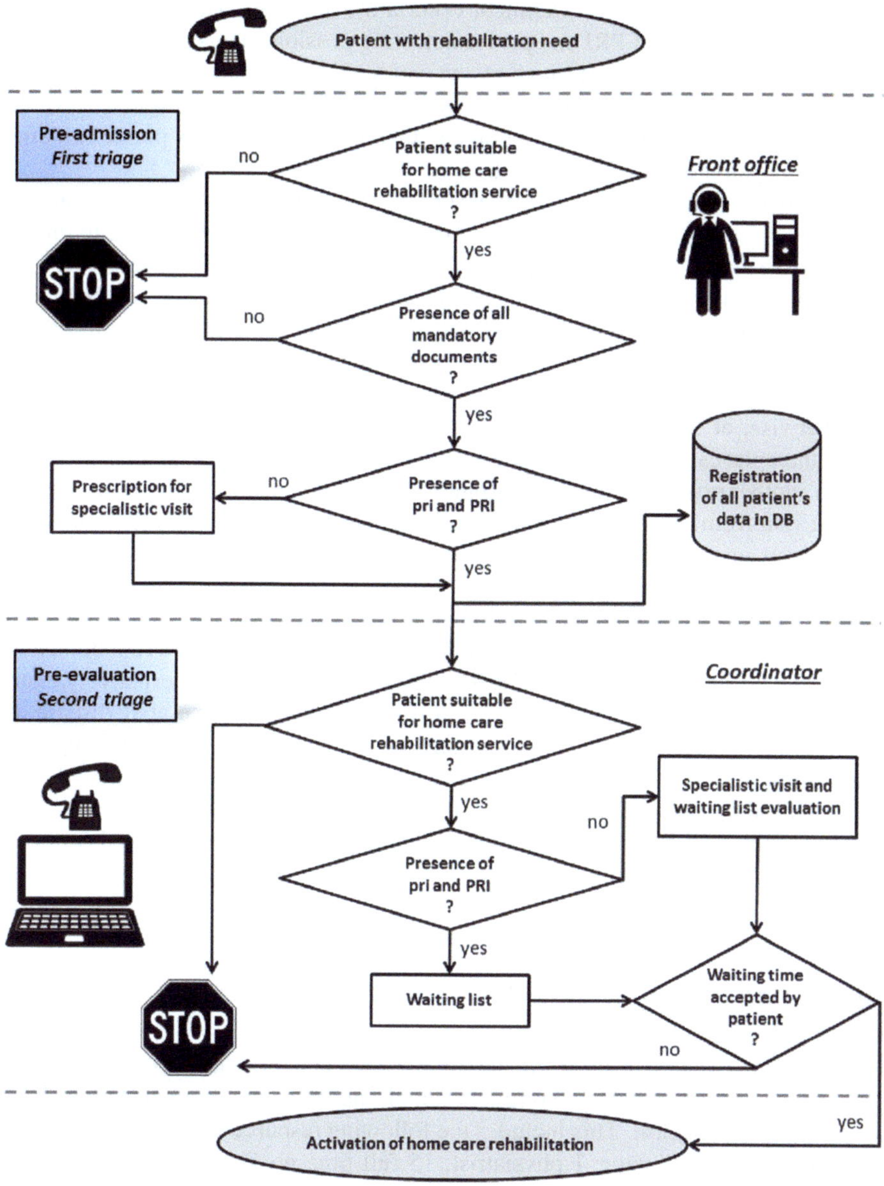

Fig. 2 Flowchart of the activities in the admission process of a new candidate patient

Finally, if the patient is eligible and when he/she has all mandatory documentation, all personal data are recorded in the database and the process goes on.

The second admission step is managed by the coordinator of the service. He/she checks all data inserted by the front office and, then, calls back the patient or the caregiver to verify that all clinical criteria are satisfied. The coordinator also rechecks the presence of PRI and pri for detecting possible errors made by the front office. If some criteria are not satisfied, the patient is discharged and the coordinator provides all information to find the best alternative. If the patient is eligible for HCR, the coordinator communicates to the patient that there will be a new phone call after checking physiotherapists' availabilities.

The coordinator checks the availability of the physiotherapists by means of phone calls. In this step, the weekly number of visits is also considered, to ensure enough time slots from the same physiotherapist at each treatment week. Then, the coordinator calls back the patient to communicate when the service can start with the first visit, or the impossibility to start within 45 days from the first phone call. If the proposed delay is not accepted by the patient or in the absence of any availability within 45 days, the activation process is stopped.

Otherwise, at the week before the first visit, the coordinator and the assigned physiotherapist call the patient to check that no problems arose (e.g., the patient has chosen another provider, has been hospitalised, or died) and fix the first visit. During the first visit from a physiotherapist, the patient is re-evaluated and the Barthel index is measured again. If the value is higher than 80 the patient is discharged; otherwise the care process can finally start.

HCR program has a nominal duration of 60 days for the majority of patients, even if an extension of one month may be required to achieve all goals, or a reduction may sometimes occur (in case the patient is hospitalised, dies, ...). Alternatively, there is no service time limit for patients who suffer from a neurodegenerative disease. In our DES model, we consider the durations extracted from a data driven empirical distribution. As for the numbers of weekly visits, they can be either extracted from a distribution or assumed fixed.

The resulting DES model has been implemented in Arena 14 (Rockwell Automation, Inc.).

Application to a Real Case and Results

In this section, we first describe the tested instances and then show the main results.

We considered the realistic configuration of one of the providers in Milan, Italy, as reference *actual* case. This includes the following resources: 2 secretaries for the front office; 1 coordinator; 1 physiatrist; 15 full time physiotherapists. Moreover, in the results presented below, we considered the actual policy observed in real providers for defining waiting lists and selecting physiotherapists (the one described in section "Discrete Event Simulation Model"). We analysed performance variations with respect to the number of physiotherapists, by adding either 1 or 2 full time physiotherapists (cases *1F* and *2F*), or either 1 or 2 part time physiotherapists (cases *1P* and *2P*).

The main (deterministic and stochastic) model parameters are reported in Table 1. In particular, two different cases are considered for the inter-arrival times of patients

Table 1 Parameters of the DES model

Type of calls	Inter-arrival time	Unit	
General calls	*Exp* (0.05)	Days	
Calls for rehabilitation service	*Exp* (0.17) or *Exp* (0.1)	Days	(a)
Calls for other services	*Exp* (0.25)	Days	

Process	Duration	Unit	
Waiting time for prescription	*Uniform* (1, 2)	Days	
Waiting time for specialistic visit for PRI and pri	*Uniform* (10, 13)	Days	(b)
Pre-evaluation	*Uniform* (0.25, 0.3)	Hours	
Medical evaluation	*Uniform* (1, 1.5)	Hours	
Call to the patient's family	*Uniform* (0.25, 0.3)	Hours	

Main patients' characteristics	Value	Unit	
Weekly visits	2	Number	(c)
Treatment duration	Data driven empirical density (mean = 95.3; st dev = 101.4)	Days	

Physiotherapists' capacities	Value	Unit	
Weekly-part time	20	Number	(d)
Weekly-full time	40	Number	

requiring HCR, i.e., an exponential law with either rate 0.17 or 0.1, meaning an expected arrival equal either to 6 or 10 per day. All of the distributions aim at reproducing realistic data and have been identified from the available historical data of providers operating in Milan, Italy.

We run all simulations until they reached the stationary behaviour, i.e., after about 3000 patients treated. Then, at the stationary behaviour, we observed in the different configurations the two main variables of interest for evaluating the quality of the admission, i.e., the waiting time and the refusal rate (the relative number of patients refused for unavailability within 45 days).

Results in Table 2 and Fig. 3 show the longest waiting times and the highest refusal rate for the reference actual case, according to its lower resources. On the contrary, for configuration *2F*, the refusal rate is always null and the shortest waiting times are observed. Similar results are obtained in configurations *2P* and *1F*, due to the capacities of part-time physiotherapists equal to half those of full-time ones. Finally, worse performances have been observed for 10 arrivals per day, as expected.

The maximum waiting time is always lower or equal to 33, while higher average values correspond to a higher refusal rate rather than to an increased maximum waiting time. Null refusal rate is observed with 6 arrivals per day, also in the *actual* configuration, meaning that the system is already able to take care of the demand; adding resources only reduces waiting times. With 10 arrivals per day, the *actual*

Table 2 Waiting time to start the treatment in days for the different configurations: minimum, average and maximum values

Configuration	Minimum	Average	Maximum
Waiting time — 6 arrivals per days			
Actual	6	23	33
1P	1	11	32
2P	1	11	32
1F	1	11	32
2F	0	0	0
Waiting time — 10 arrivals per days			
Actual	20	31	33
1P	15	29	33
2P	17	27	33
1F	11	28	33
2F	1	25	33

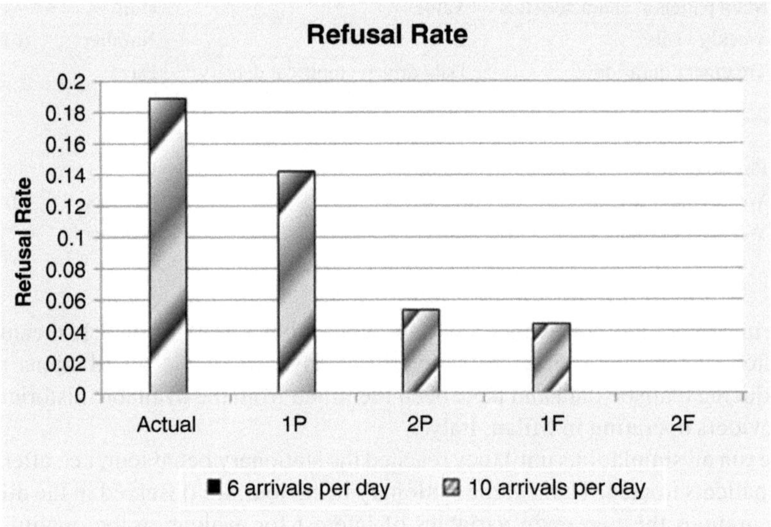

Fig. 3 Refusal rate for the different configurations (always null with 6 arrivals per day)

configuration shows a relevant refusal rate, equal to 0.19 (i.e., about 1 eligible patient over 5 is refused), and longer waiting times. Adding resources considerably reduces the refusal rate and, secondarily, also reduces waiting times, even if the impact on this second variable is lower; a larger number of added resources is required to get a higher impact also on waiting times.

Similar considerations can be derived by taking into account the number of patients in the waiting list over the simulated days.

These results are obtained with the actual policy; modifying the admission policy, e.g., with a selection of few patients to exclude for improving the service of the remaining ones, could improve system performance.

Discussion and Conclusions

We aim at developing a DES framework for analysing and optimising admission policies and waiting list management strategies in HC services. The goal is to provide a quantitative tool for evaluating performances and optimising resources in front of the qualitative evaluation usually considered in several providers. In this paper, we start our analysis proposing and testing a DES model for managing the admission process of new acute patients to a HCR service and the related waiting lists. Such task is of fundamental importance in the HC services, in particular when patients require a fast beginning of the care pathway to get the desired benefits from the therapy (e.g., in the HCR service for acute patients analysed in this paper). Results show the ability of the model to fit the real admission process and to provide quantitative information. Finally, the model also represents an useful tool for analysing the opportunity cost of increasing the resources, providing an evaluation of the impact of new resources on refusal rate reduction and the number of admitted patients.

Future work will be conducted in evaluating alternative policies by exploiting the developed framework. Moreover, the division of the territory covered by the provider into districts will be added, and operators' unavailabilities (e.g., holidays or illness) will be included.

References

Bertels, S., Fahle, T.: A hybrid setup for a hybrid scenario: combining heuristics for the home health care problem. Comput. Oper. Res. **33**, 2866–2890 (2006)

Carello, G., Lanzarone, E.: A cardinality-constrained robust model for the assignment problem in home care services. Eur. J. Oper. Res. **236**, 748–762 (2014)

Collin, C., Wade, D.T., Davies, S., Horne, V.: The Barthel ADL index: a reliability study. Int. Disabil. Stud. **10**, 61–63 (1988)

Deliberation IX/3851 of Region Lombardia, Italy (July, 25, 2012)

Eveborn, P., Flisberg, P., Ronnqvist, M.: LAPS CARE - an operational system for staff planning of home care. Eur. J. Oper. Res. **171**, 962–976 (2006)

Günal, M.M., Pidd, M.: Discrete event simulation for performance modelling in health care: a review of the literature. J. Simul. **4**, 42–51 (2010)

Hertz, A., Lahrichi, N.: A patient assignment algorithm for home care services. J. Oper. Res. Soc. **60**, 481–495 (2009)

Jacobson, S.H., Hall, S.N., Swisher, J.R.: Discrete-event simulation of health care systems. In: Patient Flow: Reducing Delay in Healthcare Delivery. International Series in Operations Research & Management Science, Springer vol. 91, pp. 211–252 (2006), http://link.springer.com/chapter/10.1007%2F978-0-387-33636-7_8

Kommer, G.J.: A waiting list model for residential care for the mentally disabled in the Netherlands. Health Care Manage. Sci. **5**, 285–290 (2002)

Lanzarone, E., Matta, A.: A cost assignment policy for home care patients. Flex. Serv. Manuf. J. **24**, 465–495 (2012)

Lanzarone, E., Matta, A.: Robust nurse-to-patient assignment in home care services to minimize overtimes under continuity of care. Oper. Res. Health Care **3**, 48–58 (2014)

Lanzarone, E., Matta, A., Sahin. E.: Operations management applied to home care services: the problem of assigning human resources to patients. IEEE Trans. Syst. Man Cybern. Part A Syst. Humans **42**, 1346–1363 (2012)
Mahoney, F.I., Barthel, D.W.: Functional evaluation: the Barthel index. Md. State Med. J. **14**, 61–65 (1965)
Meiland, F.J., Danse, J.A., Hoos, A.M., Wendte, J.F., Gunning-Schepers, L.J.: The use of the waiting list in a fair selection of patients for nursing home care. Health Policy **38**, 1–11 (1996)
Mitchell, C., Walker, J., Walters, S., Morgan, A.B., Binns, T., Mathers, N.: Costs and effectiveness of pre- and post-operative home physiotherapy for total knee replacement: randomized controlled trial. J. Eval. Clin. Pract. **11**, 283–292 (2005)
Regional Decree of Lombardia, Italy, no. 6032 (July, 06, 2012)
Sainsbury, A., Seebass, G., Bansal, A., Young, J.B.: Reliability of the Barthel index when used with older people. Age Ageing **34**, 228–232 (2005)
Wijkstra, P.J., Ten Vergert, E.M., Van Altena, R., Otten, V., Kraan, J., Postma, D.S., Koëter, G.H.: Long term benefits of rehabilitation at home on quality of life and exercise tolerance in patients with chronic obstructive pulmonary disease. Thorax **50**, 824–828 (1995)
Wiles, C.M., Newcombe, R.G., Fuller, K.J., Shaw, S., Furnival-Doran, J., Pickersgill, T.P., Morgan, A.: Controlled randomised crossover trial of the effects of physiotherapy on mobility in chronic multiple sclerosis. J. Neurol. Neurosurg. Psychiatry **70**, 174–179 (2001)
Yalçındağ, S., Matta, A., Sahin, E.: Operator assignment and routing problems in home health care services. In: Proceedings of the IEEE International Conference on Automation Science and Engineering (CASE), pp. 329–334 (2012)
Yalçındağ, S., Matta, A., Sahin, E., Shanthikumar, J.G.: A two-stage approach for solving assignment and routing problems in home health care services. In: Proceedings of the International Conference on Health Care Systems Engineering. Springer Proceedings in Mathematics & Statistics, vol. 61, pp. 47–59 (2014)

Ambulance Location Problem with Stochastic Call Arrivals Under Nearest Available Dispatching Policy

Inkyung Sung and Taesik Lee

Introduction

An ambulance location problem has been extensively studied since as early as 1970s (Church and ReVelle 1974; Toregas et al. 1971). The problem is to determine the locations of ambulances to provide maximum coverage to potential demand sites. Locations of available ambulances are a major factor to determine the response times to arriving calls. Brotcorne et al. (2003); Farahani et al. (2012); Li et al. (2011); Owen and Daskin (1998); ReVelle and Eiselt (2005) provide a comprehensive review of location problems in emergency medical service (EMS) systems.

Ambulance location problems are often formulated as a covering problem. A demand site is considered covered if it can be reached from an ambulance station within a time standard. Then, the problem finds optimal number and locations for ambulances so that the sum of covered demand sites is maximized. The classic ambulance location problems model the coverage as deterministic. It assumes that ambulances are always available to respond to emergency calls. On the other hand, more recent models incorporate randomness in ambulance's availability. These models often use the concept of busy fraction of an ambulance—probability for being unavailable to respond to a call.

Our study has been motivated by the fact that an ambulance dispatching policy is an important factor affecting ambulance's availability. An ambulance dispatching policy determines which of the ambulances available at the moment is sent to serve an incoming call. A choice made for the current call determines the available ambulances and their coverage for the next arriving call. This implies that there

I. Sung • T. Lee (✉)
KAIST, 291 Daehak-ro, Yuseong-gu, Daejeon 305-701, Republic of Korea
e-mail: inkyung@kaist.ac.kr; taesik.lee@kaist.edu

A. Matta et al. (eds.), *Health Care Systems Engineering for Scientists and Practitioners*, Springer Proceedings in Mathematics & Statistics 169,
DOI 10.1007/978-3-319-35132-2_10

is an interaction effect between the location decision and dispatching policy; an optimal location solution under one dispatching policy may not be the optimal solution for another policy. Therefore ambulance locations should be determined while considering an ambulance dispatching policy. We construct our model for a specific dispatching policy to explicitly incorporate the effect of a dispatching policy on ambulance location solution. Given a temporal sequence of call arrivals, which is sampled from real data, the model simultaneously determines ambulance locations and call assignments to ambulances under the dispatching policy.

Our model is also designed to take into account variations in call arrivals. Most of the probabilistic models in the ambulance location literature assume stationary call arrivals and use an average call arrival rate in the model. However, the actual EMS call data that the volume of call arrivals varies significantly over the course of a day, weekdays vs. weekends, and between seasons (e.g., Matteson et al. 2011). To properly represent the variations in call arrivals, we take a stochastic programming approach and incorporate the uncertainty of call arrivals into location decisions.

Problem Formulations and Scenario Decomposition Implementation

We develop a solution algorithm for an ambulance location problem that takes into account an ambulance dispatching policy and variations in EMS call arrivals. For this, we apply a stochastic programming approach. Stochastic programming is a framework for modeling an optimization problem with two types of decision variables, *here-and-now* and *recourse*. A here-and-now decision is a proactive and planning decision that should be made before observing specific outcomes (e.g., production cost or future demands). A recourse decision is made in reaction against the observations on the outcome, and the here-and-now decision made earlier is adapted accordingly. In general, a recourse decision depends upon a here-and-now decision. With the decision structure, stochastic programming enables to consider uncertainties in the outcomes and derive a solution that is robust to the uncertainties. Stochastic programming uses a set of scenarios, i.e., possible futures, and derives solutions that perform well across all scenarios. Scenarios is a realization of the uncertainties.

In our algorithm, an ambulance location decision corresponds to the here-and-now decision, and a dispatching is the recourse decision. A scenario in our problem is defined by a sequence of call arrivals. We determine ambulance locations before observing call arrivals under a particular scenario. Then after observing actual call arrivals, ambulance dispatching decisions are made given the ambulance location decision. By taking a stochastic programming approach, we derive an ambulance location solution that performs well across all possible scenarios.

Scenario Decomposition for Stochastic Programming

Let $\boldsymbol{\xi}$ denote a random vector for call arrivals with a support Ξ and known distribution P. We assume that $\boldsymbol{\xi}$ has a finite support, and there are N realizations $\boldsymbol{\xi}^r$, $r \in \{1, \ldots, N\}$. A realization of the random vector is referred to as a *scenario*. We consider a stochastic program for an ambulance location problem:

$$\max \left\{ \sum_{r=1}^{N} p_r \cdot f(\mathbf{x}, \boldsymbol{\xi}^r) : \mathbf{x} \in X \subseteq \mathbb{Z}^+ \right\}, \tag{1}$$

where $\mathbf{x}$ is a decision vector for ambulance locations, p_r is the probability of scenario r, and $f(\mathbf{x}, \boldsymbol{\xi}^r)$ is the sum of covered demands by solution $\mathbf{x}$ under scenario $\boldsymbol{\xi}^r$.

One of the approaches to solve (1) is *scenario decomposition*. The main idea of scenario decomposition is to decompose the main problem (1) into N sub-problems by maintaining an individual copy of here-and-now decision variable for each scenario, $\mathbf{x}^r$. Because location solutions should not depend on a scenario, *non-anticipativity* constraint is imposed to require $\mathbf{x}^1 = \cdots = \mathbf{x}^N$. With the non-anticipativity constraint represented by the equality $\sum_{r=1}^{N} \mathbf{A}^r\mathbf{x}^r = 0$, (1) can be rewritten as

$$\max \left\{ \sum_{r=1}^{N} f_r(\mathbf{x}^r) : \mathbf{x}^r \in X \ \forall r, \ \sum_{r=1}^{N} \mathbf{A}^r\mathbf{x}^r = 0 \right\}, \tag{2}$$

where $f_r(\mathbf{x}^r) = p_r \cdot f(\mathbf{x}^r, \boldsymbol{\xi}^r)$. Unfortunately, it is difficult to incorporate the non-anticipativity constraint to the sub-problems. To overcome the difficulty, Lagrangian relaxation is applied to (2) (Fisher 2004). By dualizing the non-anticipativity constraint, the Lagrangian dual of (2) is obtained as follows:

$$\min_{\lambda} \left\{ z_D = \sum_{r=1}^{N} \max\{f_r(\mathbf{x}^r) + \boldsymbol{\lambda}\mathbf{A}^r\mathbf{x}^r : \mathbf{x}^r \in X \ \forall r\} \right\}, \tag{3}$$

where $\boldsymbol{\lambda}$ is a dual vector. This provides an upper bound for (2). By choosing $\boldsymbol{\lambda}$ such that $\mathbf{x}^r$ is identical for all $r \in \{1, \ldots, N\}$, we can find an optimal solution of (2) (Carøe and Schultz 1999). To solve (3), we implement a simple algorithm, following an algorithm proposed by Ahmed (2013). The algorithm is shown in Algorithm 1.

Algorithm 1 produces candidate solutions of the sub-problems for each scenario and calculates upper bound by the sum of the Lagrangian objective functions of the candidate solutions. A lower bound is also calculated by evaluating the original objective values of the candidate solutions. After obtaining the candidate solutions for each scenario, Algorithm 1 updates $\boldsymbol{\lambda}$ such that the upper bound is tightened. Then this procedure is repeated until the gap between lower and upper bound is close enough or the number of iterations reaches a certain threshold value.

Algorithm 1 Scenario decomposition

```
UB ← ∞, LB ← −∞, x* ← ∅, n_I ← 0, λ ← 0
while UB − LB > ε and n_I < n_max do
    n_I ← n_I + 1
    for r = 1 to N do
        solve max{f_r(x) + λA^r x}
        let v^r be the optimal value and x^r be an optimal solution
        u ← 0
        for r' = 1 to N do
            compute f_{r'}(x^r) and set u ← u + f_{r'}(x^r)
        end for
        if LB < u then
            LB ← u, x* ← x^r
        end if
    end for
    UB ← Σ_{r=1}^{N} v^r
    update λ
end while
```

To update $\boldsymbol{\lambda}$, we use a sub-gradient method. Given $\boldsymbol{\lambda}'$, the sub-gradient method calculates the gradient of z_D at $\boldsymbol{\lambda}'$. The gradient at $\boldsymbol{\lambda}'$ is given by $\sum_{r=1}^{N} \mathbf{A}^r \mathbf{x}^r$, where $\mathbf{x}^r$ is the solution to $\max\{f_r(\mathbf{x}) + \boldsymbol{\lambda}'\mathbf{A}^r\mathbf{x}\}$. Then the method updates the current $\boldsymbol{\lambda}'$ by using the gradient such that (3) can be minimized.

Modeling for the Ambulance Location Problem Given a Single Scenario

In the solution approach described in section "Scenario Decomposition for Stochastic Programming", we need to solve for each scenario the following problem:

$$\max\{f_r(\mathbf{x}) + \boldsymbol{\lambda}\mathbf{A}^r\mathbf{x}\} \tag{4}$$

We formulate this problem as an integer program. Recall that we want to incorporate the effect from a dispatching policy into the location decision. This is achieved by introducing a constraint to ensure that ambulances are assigned to arriving calls based on a chosen dispatching policy. In a sense, it is a location-routing problem with a restriction that a routing is determined once a location decision is given. For our problem, we choose the "nearest available" dispatching policy, which sends the closest available ambulance to an arriving call. This policy is a common practice found in many EMS call centers in Korea.

In the model, we have one principal decision variable and two auxiliary variables:

- integer variable x_t^j that indicates the number of available ambulances at station j at time t

- (auxiliary) binary variable y_d^j that indicates whether a call d is serviced by ambulance located at station j
- (auxiliary) binary variable z_t^j that indicates whether there is any ambulance available at station j right after an ambulance is dispatched at time t

Note that x_0^j defines the ambulance locations at the beginning of a planning horizon, hence it is the location solution for our problem. Also note that y_d^j specifies ambulance dispatching decisions for call d, and it is determined by the nearest available dispatching policy assumed in our model.

Before presenting the full formulation, the notation used in the model is summarized in Table 1.

Herc, we rewrite (4) by using W_d^j and a^r. The first term in (4) is the number of covered demands. W_d^j specifies whether demand location d can be covered by station j or not. Using W_d^j, $f_r(\mathbf{x})$ can be written as $\sum_{d \in D} \sum_{j \in V} W_d^j \cdot y_d^j$. The second term in (4) is related to the non-anticipativity constraint. We set the constraint as $(N-1)\mathbf{x}^1 = \mathbf{x}^2 + \cdots + \mathbf{x}^N$, and the second term in (4) can be written as $\lambda^j a^r \sum_{j \in V} x_0^j$, where λ^j is jth element of $|V|$ dimensional vector $\boldsymbol{\lambda}$, and $a^r = N - 1$ if $r = 1$, otherwise, -1. Then, for given $\boldsymbol{\lambda}$, the objective function can be written as

Table 1 Summary of notation

Symbol	Definition
i, j	Ambulance station index
t	Time index
d	Call index
V	Set of candidate ambulance locations
T	Final time horizon
D	Set of calls occurred during a planning horizon
q	The number of maximum ambulances
a_d	Arrival time of call d
N_t	The number of calls during time interval t
τ_d^j	Distance, measured in time, between call d and station j
R_d^j	Turn around time of call d by an ambulance in station j
S	Time standard for coverage
M	Sufficiently large number
A_t	Set of calls arriving at time t
B_t^j	Set of calls for station j satisfying the condition, $a_d + R_d^j = t$
W_d^j	Constant: 1 if $\tau_d^j \leq S$; 0, otherwise
x_t^j	Variable: the number of available ambulances at station j at time t
y_d^j	Variable: 1 if ambulance j dispatched to call d; 0, otherwise
z_t^j	Variable: 1 if at least one ambulance is available at station j right after time t; 0, otherwise

$$\max \sum_{d \in D} \sum_{j \in V} W_d^j \cdot y_d^j + \lambda^r a^r \sum_{j \in V} x_0^j, \tag{5}$$

The constraints of the model are constructed to impose the nearest–available dispatching policy. Before describing the constraints we introduce two assumptions. First, ambulances on its way back to its home station are not available for service until it returns to the station. Second, if a call arrives and all ambulances are busy at the moment, the call is either lost or served from other EMS systems. These assumptions can be justified by the fact that probabilities of the events are very small. In addition, standard practice for operating ambulances in Korea is to return its home station after serving a call in order to get ready for next call arrivals. With these assumptions, the objective function (5) is solved subject to the following set of constraints:

$$\sum_{j \in V} x_0^j = q \tag{6}$$

$$\sum_{j \in V} y_d^j \leq 1 \qquad \forall d \in D \tag{7}$$

$$x_t^j = x_{t-1}^j - \sum_{d \in A_{t-1}} y_d^j + \sum_{d \in B_t^j} y_d^j \qquad \forall t \geq 1, j \in V \tag{8}$$

$$\sum_{d \in A_t} \sum_{j \in V} y_d^j = \min\left(N_t, \sum_{j \in V} x_t^j\right) \qquad \forall t \leq T \tag{9}$$

$$x_t^j - \sum_{d \in A_t} y_d^j \geq z_t^j \qquad \forall t \leq T, j \in V \tag{10}$$

$$x_t^j - \sum_{d \in A_t} y_d^j \leq M \cdot z_t^j \qquad \forall t \leq T, j \in V \tag{11}$$

$$y_d^j \cdot \tau_d^j \leq M(1 - z_t^i) + z_t^i \cdot \tau_d^j \qquad t = a_d, \forall d \in D,\ i, j \in V, i \neq j \tag{12}$$

$$x_t^j \in \mathbb{Z}^+ \qquad \forall t \leq T, j \in V \tag{13}$$

$$y_d^j \in \{0, 1\} \qquad \forall d \in D, j \in V \tag{14}$$

$$z_t^j \in \{0, 1\} \qquad \forall t \leq T, j \in V \tag{15}$$

Constraint (6) limits the total number of ambulances to be located at q. Constraint (7) ensures that at most one ambulance is dispatched to serve a call. Constraint (8) determines the number of available ambulances at station j at the beginning of time interval t, x_t^j. It is computed by subtracting ambulances dispatched from station j to calls during time period $t - 1$ and adding ambulances that are returning to the station at the beginning of time interval t. Constraint (9) ensures the number of ambulances dispatched during time interval t equals to either the number of calls during time interval t or the number of available ambulances at the beginning of the time interval. By constraint (10) and (11), z_t^j becomes one if station j has at least one ambulance at time $t + \epsilon$, where $t + \epsilon$ is the time right after a dispatching decision at station j is made. Then, constraint (12) ensures ambulances are assigned based on the nearest available dispatching policy. It states that travel time for call d from station j is smaller than any other station at which ambulances are available. It should be noted that other types of dispatching policy based on priority (e.g., regionalized response Aboueljinane et al. 2013) can be modeled in the same way by changing the travel time to corresponding cost measures.

In Algorithm 1, we need to solve this sub-problem for all scenarios per each iteration, for a large number of iterations. Therefore it is important to quickly solve the sub-problems to make Algorithm 1 computationally efficient. Unfortunately, our initial tests show that commercial LP solvers—we used CPLEX—cannot solve the sub-problems fast enough, and we decide to develop a meta-heuristic algorithm to quickly obtain near optimal solutions. In this study, we use Variable Neighborhood Search (VNS).

In VNS, we first define several neighborhood structures, $\mathcal{N}_k(k = 1, \ldots, k_{max})$. A neighborhood structure specifies distance between two candidate solutions, which is used to identify neighbors for current solution $\mathbf{x}$. VNS uses several neighborhood structures to avoid local optima by exploring a large solution space, including distant neighborhood of a current solution. The solution structure for location problem is simple and easy to measure the distance between two feasible solutions. For these reasons, VNS algorithm can be readily implemented for location problems. We follow the basic structure of VNS described in Hansen and Mladenovic (2001), and it is depicted in Algorithm 2.

In Algorithm 2, $Shaking(\mathbf{x}, k)$ randomly generates a solution $\mathbf{x}'$ from the kth neighborhood of $\mathbf{x}$. After the solution $\mathbf{x}'$ is obtained, a local search method $LocalSearch(\mathbf{x}')$ is applied to improve solution $\mathbf{x}'$. In our implementation, we search all neighborhoods of $\mathbf{x}'$ in $\mathcal{N}_2$ and return the best solution among them. Then, the resulting solution $\mathbf{x}''$ is accepted if $\mathbf{x}''$ is better than current incumbent solution.

Algorithm 2 Basic VNS

```
Select the set of neighborhood structures N_k, k = 1, ..., k_max
Generate initial solution, x
repeat
    k ← 1
    repeat
        x' ← Shaking(x, k)
        x'' ← LocalSearch(x')
        if accept(x'') then
            x ← x''
            k ← 1
        else
            k ← k + 1
        end if
    until k = k_max
until stopping condition is met
```

For the implementation of VNS, we define the set of neighborhood structures as follows:

$$\mathcal{N}_k(\mathbf{x}) = \{\mathbf{x}' : |\mathbf{x} \setminus \mathbf{x}'| = |\mathbf{x}' \setminus \mathbf{x}| = k\}.$$

If a location solution $\mathbf{x}'$ differs from $\mathbf{x}$ in k locations, i.e., $|\mathbf{x} \setminus \mathbf{x}'| = |\mathbf{x}' \setminus \mathbf{x}| = k$, then $\mathbf{x}'$ belongs to a neighborhood of $\mathbf{x}$ in neighborhood structure $\mathcal{N}_k$.

For an initial solution to feed to the VNS algorithm, we solve the integer program (5) without considering the constraints for the nearest available dispatching policy, (9)–(12), (15). That is, we solve for the objective function (5) with a partial set of constraints (6)–(8), (13)–(14). In doing so, we need a new constraint to replace the original constraint (9):

$$\sum_{d \in A_t} y_d^j \leq x_t^j \qquad \forall t \leq T, j \in V.$$

This is to ensure the number of ambulances dispatched during time interval t do not exceed the number of available ambulances available at the beginning of the interval t.

Results

We test our solution algorithm by using EMS call data for the city of Daejeon in Korea. EMS log data for the month of January of 2010 is used to generate a set of scenarios. The algorithm determines ambulance locations based on the scenarios. Then the solution is evaluated by using the call data from February of 2010. For evaluation, we measure the percentage of the calls to which an ambulance arrives within 10 min. For call arrivals, we used the actual data and ambulances are dispatched following the nearest available dispatching policy. As a comparison, we obtain ambulance location solutions by using MALP II (ReVelle and Hogan 1989) and BACOP2 (Hogan and ReVelle 1986). MALP II is a probabilistic ambulance location model to maximize the number of demands covered. Incorporated in the model is the availability constraint, which requires a demand point be covered by multiple number of ambulances. Workload for ambulances determine how many ambulances should cover a demand point to ensure certain level of ambulance availability (60 % in this experiment). BACOP2 also aims to address the availability of ambulances, but it does so by requiring a fixed number of ambulances cover a demand point.

In the experiment, we vary the number of ambulances q to locate, and compare the performance of location solutions by the three approaches. Figure 1 shows the results for $q = 3, 5, 7, 10$.

In Fig. 1, we observe that our algorithm performs better than the other location models. In comparison with MALP II model, when the number of ambulances is

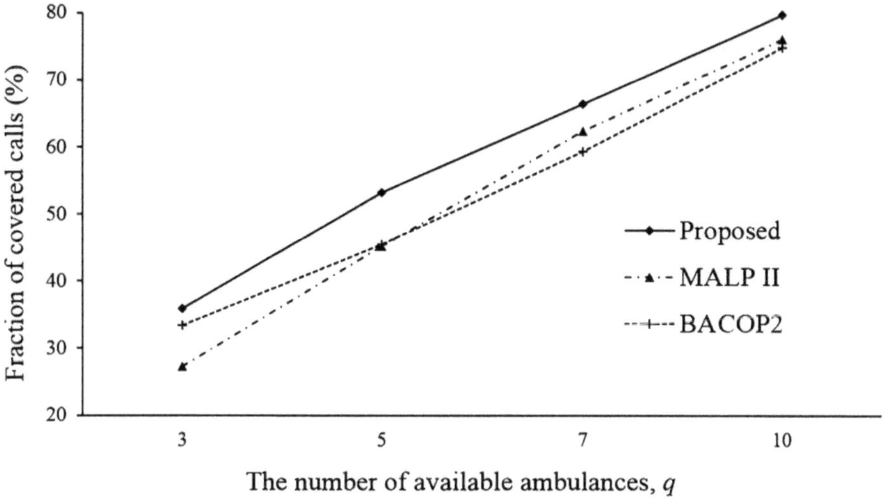

Fig. 1 Fraction of calls served by an ambulance within 10 min

low, the improvements by the proposed algorithm is significant. MALP II model tends to collocate ambulances in order to satisfy the availability constraint, leaving a large number of demands left uncovered. More importantly, it should be noted that MALP II uses busy fraction, which is estimated as an average value and ignores temporal variations in call arrivals. This possibly makes the estimated busy fraction an over-estimation for the night hours and under-estimation for the day hours. The performance gap between the proposed algorithm and BACOP2 increases as the number of ambulances is high. BACOP2 model tends to spread out ambulances in order to maximize the deterministic coverage. While such strategy seems appropriate when the number of ambulances is low, an approach that takes into account the availability of ambulances (the proposed approach and MALP II in this case) offers a larger benefit as the number of ambulances increases.

Conclusion

In this paper, we develop a model and a solution algorithm to locate ambulances. In particular, our model addresses two key factors in ambulance location decisions: a dispatching policy and temporal variations in call arrivals. The novelties of our model are (1) it explicitly describes an ambulance dispatching policy in an ambulance location problem so that the interaction between two decisions, i.e., ambulance dispatching and locations, is considered (2) our model allows us to consider the temporal variations in call arrivals which allows to incorporate ambulance availability in a more precise fashion than classical probabilistic location models.

Specifically, we model this ambulance location problem as an integer program with the constraints for the nearest available dispatching policy. We applied stochastic programming to incorporate various call arrival patterns. To obtain solutions, we implement a scenario decomposition approach which separately solves sub-problems for each scenario while maintaining the non-anticipativity. The solutions of the sub-problems are obtained by using a VNS method, a meta-heuristic algorithm. The experiments demonstrate that by considering the ambulance dispatching policy and temporal variations of call arrivals, it delivers superior performance compared with some of the classic location models.

Acknowledgements This research was supported by a grant "research and development of modeling and simulating the rescues, the transfer, and the treatment of disaster victims" (nema-md-2013-36) from the man-made disaster prevention research center, Ministry of Public Safety and Security.

References

Aboueljinane, L., Sahin E., Jemai Z.: A review on simulation models applied to emergency medical service operations. Comput. Ind. Eng. **66**, 734–750 (2013)

Ahmed, S.: A scenario decomposition algorithm for 0–1 stochastic programs. Oper. Res. Lett. **41**, 565–569 (2013)

Brotcorne, L., Laporte G., Semet, F.: Ambulance location and relocation models. Eur. J. Oper. Res. **147**, 451–463 (2003)

Carøe, C.C., Schultz, R.: Dual decomposition in stochastic integer programming. Oper. Res. Lett. **24**, 37–45 (1999)

Church, R., ReVelle, C.: The maximal covering location problem. Pap. Reg. Sci. **32**, 101–118 (1974)

Farahani, R.Z., Asgari, N., Heidari, N., Hosseininia, M., Goh, M.: Covering problems in facility location: a review. Comput. Ind. Eng. **62**, 368–407 (2012)

Fisher, M.L.: The lagrangian relaxation method for solving integer programming problems. Manag. Sci. **50**, 1861–1871 (2004)

Hansen, P., Mladenovic, N.: Variable neighborhood search: principles and applications. Eur. J. Oper. Res. **130**, 449–467 (2001)

Hogan, K., ReVelle, C.: Concepts and applications of backup coverage. Manag. Sci. **32**, 1434–1444 (1986)

Li, X., Zhao, Z., Zhu, X., Wyatt, T.: Covering models and optimization techniques for emergency response facility location and planning: a review. Math. Methods Oper. Res. **74**, 281–310 (2011)

Matteson, D.S., McLean, M.W., Woodard, D.B., Henderson, S.G.: Forecasting emergency medical service call arrival rates. Ann. Appl. Stat. **5**, 1379–1406 (2011)

Owen, S.H., Daskin, M.S.: Strategic facility location: a review. Eur. J. Oper. Res. **111**, 423–447 (1998)

ReVelle, C.S., Eiselt, H.A: Location analysis: a synthesis and survey. Eur. J. Oper. Res. **165**, 1–19 (2005)

ReVelle, C., Hogan, K.: The maximum availability location problem. Transp. Sci. **23**, 192–200 (1989)

Toregas, C., Swain, R., ReVelle, C., Bergman, L.: The location of emergency service facilities. Oper. Res. **19**, 1363–1373 (1971)

Approach to Clustering Clinical Departments

Alexander Hübner, Manuel Walther, and Heinrich Kuhn

Introduction

Medical advancements and a shift in demographics have been putting pressure on the healthcare system as costs have to be kept low. Hospital resources in particular have to be used as efficiently as possible while maintaining a sufficient level in patient care and optimizing workload for medical staff. In this context it is important to keep bed occupancy levels as high and stable as possible while minimizing the risk of not having enough beds. Theoretically, this could be done by scheduling elective patients in such a way that occupancy is leveled over time. However, due to stochastic effects in emergency arrivals, length of stay of patients as well as constraints in operations room planning, the actual bed requirements for each clinical department vary significantly. Secondly, next to having weekly seasonalities, certain departments exhibit distinct changes in occupancy levels throughout the year. For example, orthopedics—a highly elective discipline—usually has reduced bed requirements during the summer months as patients are on holidays. A practical solution to this problem is to pool respective ward capacities for a compatible set of departments to provide a balancing effect with regard to overall bed occupancy.

In this paper, we present a conceptual background and decision model designed to combine departments into clusters and assign existing ward capacities to

A. Hübner (✉)
Department of Operations, Catholic University of Eichstätt-Ingolstadt, Auf der Schanz 49, 85049 Ingolstadt, Germany
e-mail: alexander.huebner@ku.de

M. Walther • H. Kuhn
Catholic University of Eichstätt-Ingolstadt, Auf der Schanz 49, 85049 Ingolstadt, Germany
e-mail: manuel.walther@ku.de; heinrich.kuhn@ku.de

A. Matta et al. (eds.), *Health Care Systems Engineering for Scientists and Practitioners*, Springer Proceedings in Mathematics & Statistics 169,
DOI 10.1007/978-3-319-35132-2_11

these clusters. Departments are combined depending on respective medical constraints, patient compatibilities, seasonality effects, and additional management and training costs, which are required when handling larger patient numbers with different medical conditions within the same cluster. We then assign actual ward capacities to these clusters to ensure the optimal size and location for every cluster. We contribute to current literature, as this problem has not been solved efficiently in terms of solution quality and computation time as well as with regard to all of the aforementioned constraints.

The remainder is organized as follows. Section "Problem Description" gives a detailed description of the problem at hand followed by an overview of related literature in section "Related Literature and Research Gap". Section "Clustering Model" then develops a clustering model for clinical departments. The final section "Discussion" summarizes first results.

Problem Description

In this section we want to elaborate on the problem at hand to offer a deeper and more comprehensive understanding of the situation. The objective of the strategic planning problem is to combine clinical departments and wards to clusters to minimize costs for beds and personnel while maintaining a predefined high service level. Wards are assigned to clusters to provide the required amount of beds for each cluster and connect the placement and size of clusters within the hospital (see Fig. 1).

Basically, the task of clustering departments can be broken down into four different parts: *medical and patient compatibility*, *occupancy level effects*, *personnel qualification and complexity management*, and *allocation of ward space*. The following paragraphs will deliver some context to this.

Medical and Patient Compatibility When patients with different medical conditions are using the same ward, it is imperative that there be no potential healthcare issues due to the close proximity of the patients. For example, immunocompromised

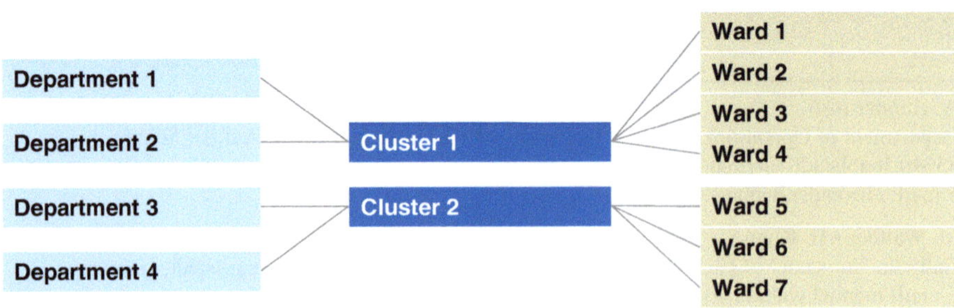

Fig. 1 Assignment of departments and wards to clusters

patients must not be exposed to infectious patients. The potential risk holds not only for shared rooms but also for entire wards as patients may freely move around and use common areas, e.g. bathrooms. Even if it were possible to confine patients to their respective rooms, thorough cleaning procedures would be needed after patient rotations to prepare the rooms for the next patients.

Patient compatibility as opposed to medical compatibility describes societal and emotional requirements which need to be met when combining different departments. This comprises differences in severity of individual medical conditions, age, and gender. For example, mixing a specialized department for geriatric dementia patients with an orthopedics department that specializes in sporting injuries might lead to tension on joint wards due to age differences and conflicting therapy plans. Also, although medically possible, children for example should not be put on the same ward as adults, even if they suffer from similar medical conditions. The same principle can be applied to strictly female departments, e.g., you would not combine an obstetric ward with a general surgical ward.

Occupancy Level Effects In general, occupancy levels of any department are a function of patient arrival rates and lengths of stay. Patient arrivals comprise emergency arrivals as well as elective patients. Generally speaking, emergency arrivals can be approximated by an exponential distribution or a Poisson process (van Essen et al. 2015), which means that the probability of a patient arriving is the same for each day and that arrivals are not controllable. Elective arrivals on the other hand are mostly plannable. However, there are certain constraints to scheduling elective patients. Depending on the severity and acuteness of one's medical condition, necessary treatments have to be conducted within a certain period of time. Moreover, elective patients requiring surgery have to be planned in accordance with Operating Room (OR) availability, which in turn depends, amongst others on personnel planning, master surgery schedule planning, and changeover times.

Theoretically, one could argue that by anticipating emergency arrivals and quickly adapting short-term elective arrivals to current occupancy levels, it is possible to completely even out occupancy over time. However, real life data we have sampled in large German hospitals suggests otherwise, due to the aforementioned constraints. Departments typically exhibit weekly seasonality effects where peak levels are reached during the week while wards are usually less crowded on weekends. This is mainly due to the fact that elective surgeries are scheduled between Monday and Friday. Note, that the actual peak day during a week differs between departments due to the above-mentioned constraints.

Naturally, these weekly peak occupancy levels need to be used to determine ward sizes while maintaining a predefined service level. As clustering departments can be considered a strategic decision which is made for several months or years it proves beneficial to combine departments in such a way that overall occupancy for the cluster is leveled (see Fig. 2).

Finally, seasonality effects are not limited to weekly occupancy fluctuations but also comprise changes in occupancy between seasons. For example, highly

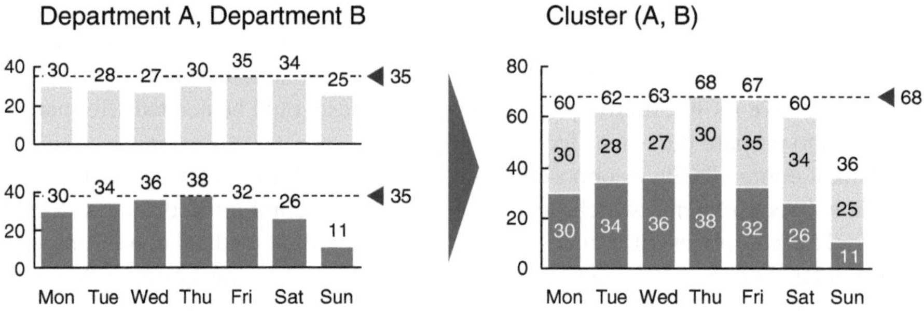

Fig. 2 Weekly seasonality

elective medical disciplines such as orthopedics show reduced patient arrivals during the summer when patients are on vacation. Thus, anticipating monthly increases or decreases in patient occurrences plays an important role in assigning and dimensioning ward capacities.

Personnel Qualification and Manageability When multiple departments have access to the same ward, every available bed can be assigned to any patient of that cluster as long as each room is occupied by only one gender at a time. As opposed to doctors who usually only visit their patients once a day and otherwise spend time in the OR or doing consultations, nursing staff are the first contact for patient needs and thus have to frequently interact with them. Therefore it would be counterproductive if they only catered to one department as their patients would then be "scattered" all over the cluster. In consequence, all nursing staff should be affiliated to specified wards where they need to be able to handle all potential patient types of that respective cluster. This means understanding and being able to act upon department-specific medical needs. Thus, additional qualification of nursing staff might be necessary when combining departments.

Furthermore, a combined resource pool for multiple departments adds substantial complexity in terms of managing elective patient arrivals, bed assignment and patient pathways. Hence, it is important to be able to handle such complexity in order to achieve an added efficiency through clustering. In general, this can only be done by providing sufficient management resources and appropriate software support to create the required transparency across multiple units.

Allocation of Ward Space Last but not least, existing ward capacities have to be assigned to clusters. Ward size, ward position, ward equipment and distance to fixed facilities are key parameters to this. The overall ward capacity assigned to any one cluster should be as close as possible to the required amount of beds needed for that cluster to ensure an efficient utilization of resources. Also, all wards assigned to a single cluster should be in close proximity to each other. This is especially important for all attending physicians as they are required to visit all of their patients and need to avoid losing time during rounds.

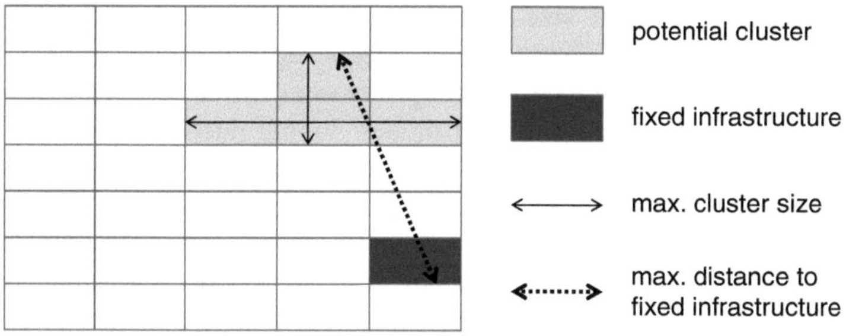

Fig. 3 Allocation of ward space

Some wards may also be equipped with specialized fixed equipment needed only by selected departments which makes it beneficial to assign them to these respective departments to forego potential conversion costs. Next to fixed equipment on wards, e.g., integrated oxygen supplies, it is also important to consider the distance between a ward and necessary fixed facilities within a hospital. For example, delivery rooms should be close to the obstetric wards and the cardiological catheter OR should be close to the cardiology department.

As stated in the previous paragraph, available ward space can be occupied by any department of a given cluster. In consequence, constraints regarding ward location are always dependent on the department within that cluster with the least degree of freedom in terms of distance to required fixed facilities (see Fig. 3).

Related Literature and Research Gap

In this section we want to give an overview over related literature and point out identified research gaps. Simply put, the problem we are investigating consists of allocating ward space to departments or groups of departments, i.e. clusters, to achieve economic benefits through pooling effects. Within this problem setting we can distinguish between modeling patient arrivals, calculating bed requirements, assigning ward space, and determining managerial and personnel-related effects. To our knowledge, current literature mostly discusses parts of this problem whereas there are only a few approaches, notably by van Essen et al. (2015) and De Bruin et al. (2010) which combine patient arrivals rates and clustering of departments.

Related Literature Among others, patient arrivals have often been modeled using classical queueing theory. For example, Gorunescu et al. (2002) describe a model using an M/Ph/c/N queue which includes a maximum queueing capacity N. They formulate a cost model that penalizes patient refusals and differentiates between waiting room (inventory) and hospital (holding) costs to determine optimal occupancy levels for a single department with given bed capacities. De Bruin et al. (2010) make use of the Erlang loss formula (a M/G/c/c queueing system) to determine optimal ward size and show that it can be beneficial to cluster departments in order to achieve leveling effects regarding joint ward occupancy. They also acknowledge that weekly occupancy rates differ between weekdays and weekends, which they separate in their model. Taking into account that admission schedules may vary strongly throughout the course of a week, Lapierre et al. (1999) use a time series model based on hourly census data to allocate beds to departments. Another approach to modeling patient arrivals is taken by Utley et al. (2003) who consider the classical Erlang loss approach disadvantageous when looking at departments with substantial shares of elective patients. They propose to use probability distribution functions for occupancy levels to support ward dimensioning decisions. When actual individual expected distributions of patient arrivals are known, a very promising approach to estimating combined occupancy level distribution functions is to convolve the deterministic data. For example, this approach has been followed by Fügener et al. (2014). In their recent paper, van Essen et al. (2015) propose a model which aims at clustering departments and assigning suitable ward space to the respective clusters. Using the Erlang loss formula, they calculate bed requirements of potential clusters. Their objective is to minimize walking distances within clusters and between clusters and fixed facilities while minimizing the maximum number of departments per cluster to avoid unnecessary complexity.

Research Gap The articles discussed above make contributions in terms of improving estimation and determination of bed capacity requirements. They further show that clustering across wards and departments allows achieving higher bed occupancy levels and ultimately achieving lower costs per patient by pooling resources. One of the key difficulties that all papers are discussing is choosing the correct approach to estimate bed requirements. Many approaches resort to the Erlang loss formula to estimate arrival times and lengths of stay to determine occupancy levels. This is problematic as the Erlang loss approach does not account for weekly or monthly seasonality effects and assumes arrival rates following a Poisson process, which is typically not the case with elective patient arrivals due to surgery schedule and personnel constraints among other reasons. Furthermore, non of the approaches presented in literature combine all the necessary aspects of clustering departments and wards as described in section "Problem Description".

Contribution In summary, our approach to clustering departments and wards contributes to current literature on three distinct aspects. First, our solution approach provides a time-efficient solution method, which finds the optimal solution for typical hospital dimensions without having to resort to heuristic solutions (see for example van Essen et al. (2015)). Second, our approach takes weekly and monthly seasonality effects into account and delivers more robust results when handling differing occupancy patterns (e.g., high elective vs. high emergency share) compared to the Erlang-Loss approach used in most publications. Third, our approach allows for high levels of customizability of department-individual constraints (e.g., service levels, location constraints) due to limitation of the solution space.

In the following section, we will develop our decision model to cluster departments and wards to extend current literature where we incorporate seasonality effects and consider personnel and management constraints as well as provide an efficient modeling and solution approach that can be applied to large hospitals and solved to optimality.

Clustering Model

This section formulates the model and associated solution approach which we used to cluster clinical departments and assign given ward capacities to the resulting clusters. To be able to solve this optimization problem in finite time, we chose a set-partitioning formulation that moves much of the required calculations into preprocessing steps. The key approach here is to rule out infeasible and impractical combinations prior to solving the model, thus enabling us to radically downsize the potential solution space while still allowing for maximum customizability of input parameters.

We use the following notation as denoted in Table 1.

$$min \sum_{c \in C} \varphi_c \cdot X_c \tag{1}$$

subject to

$$\sum_{c \in C} Q_{cd} \cdot X_c = 1 \qquad \forall\, d \in D \tag{2}$$

$$Y_{cw} \leq X_c \qquad \forall\, c \in C; \forall\, w \in W \tag{3}$$

$$\sum_{c \in C} Y_{cw} \leq 1 \qquad \forall\, w \in W \tag{4}$$

$$Z_{wv} \geq Y_{cw} + Y_{cv} - 1 \qquad \forall\, c \in C; \quad \forall\, w, v \in W \tag{5}$$

$$r_{wv} \cdot Z_{wv} \leq \tilde{r} \qquad \forall\, w, v \in W \tag{6}$$

$$h_{cw} \cdot Y_{cw} \leq \widetilde{h_c} \qquad \forall\, c \in C; \forall\, w \in W \tag{7}$$

$$X_c \cdot b_c \leq \sum_{w \in W} Y_{cw} \cdot a_w \qquad \forall\, c \in C \tag{8}$$

The objective function (1) of the model minimizes the overall cost associated to the chosen cluster combination. For every potential cluster c, the cost parameter φ_c denotes the sum of the cost of beds to be held available to satisfy a given service level and the additional management and qualification cost to operate the cluster. Constraint (2) ensures that each department $d \in D$ is covered exactly once. We use a set partitioning approach where Q_{cd} equals 1 if department d is

Table 1 Notation

Sets	
$c \in C$	Possible clusters (department combinations)
$d \in D$	Departments in hospital
$w, v \in W$	Wards in hospital
Parameters	
φ_c	Accumulated cost per cluster c including cost of required beds to ensure a given service level as well as additional personnel qualification and management cost to operate the cluster
a_w	Available beds per ward w
b_c	Beds required to meet predefined blocking probability for cluster c
h_{cw}	Distance between ward w and relevant fixed facilities (e.g. OR) for cluster c
$\widetilde{h}_c$	Maximum allowed distance between any ward w of any one cluster c to the its relevant fixed facilities
Q_{cd}	Matrix to initialize set partitioning problem; $Q_{cd} = 1$ if department d is assigned to cluster c; $Q_{cd} = 0$ otherwise
r_{wv}	Distance between wards w and v
$\tilde{r}$	Maximum allowed distance between any two wards w, v within any one cluster
Decision variables	
X_c	$X_c = 1$ if cluster c is selected; $X_c = 0$ otherwise
Y_{cw}	$Y_{cw} = 1$ if combination of cluster c and ward w is selected; $Y_{cw} = 0$ otherwise
Z_{wv}	$Z_{wv} = 1$ if wards w and v are assigned to any one cluster; $Z_{wv} = 0$ otherwise

covered by cluster c and 0 otherwise. Constraint (3) denotes that wards $w \in W$ are only assigned to clusters $c \in C$ that are chosen or activated within the objective function. The decision variable $Y_{cw} = 1$ if ward w is assigned to cluster c and 0 otherwise. Further, constraint (4) ensures that any ward w can only be connected to one cluster c to avoid double booking of ward capacity. Constraints (5) and (6) are used to restrict walking distances between any two wards $w, v \in W$ of one cluster c where $\tilde{r}$ denotes the maximum distance between the farthest points within said cluster. Here, the auxiliary decision variable Z_{wv} is introduced which equals 1, if two wards w and v are allowed to be assigned to any one cluster c and 0 otherwise. Note, that Z_{wv} does not indicate that two wards are actually in the same cluster but constitutes a boundary function for Y_{cw}. Constraint (7) limits the distance of any ward w within a cluster c to necessary fixed infrastructure where $\widetilde{h}_c$ represents an individual threshold for each cluster c depending on the underlying department combination. Finally, to assign actual ward capacities, the amount of beds needed to maintain a predefined service level per cluster $X_b \cdot b_c$ should always be equal or less than the amount of beds available within the assigned wards $\sum_{w \in W} Y_{cw} \cdot a_w$ where a_w depicts the number of beds available in each ward w [Constraint (8)].

Solution Approach and First Results The model has been implemented in Delphi for precalculations. The precalculation contains various steps to determine potential cluster combinations from a medical and patient compatibility perspective, calculate bed requirements per cluster, calculate cluster costs, and determine cluster size and location constraints. Potential department combinations, i.e., clusters, are based on expert assessments regarding medical and patient compatibility (see section "Problem Description") for all department pairs. Bed requirements for clusters are calculated by convolving expected occupancy distributions of the departments within that cluster. Specifically, we convolve daily distributions individually to assure statistical independence. The actual number of beds needed for any given cluster is then determined by applying the highest predefined service level of all included departments to the expected occupancy distribution of said cluster. Note, that clusters with strongly heterogeneous service levels are therefore penalized. Ultimately, clustering departments depicts a trade-off where on one side you achieve occupancy balancing effects by combining ward capacities. In this context, the cost for providing an additional bed includes all related cost to adding the corresponding capacity, i.e., infrastructure (extra rooms), personnel, OR-capacity and so forth. On the other side, clustering departments requires additional personnel qualification and management costs which are also determined based on expert assessments. To assign clusters to wards, size and location constraints have to be considered (see section "Problem Description" for details). These location constraints are predefined for each department-ward combination and are used as foundation to determine which wards may be used by which clusters. Note, that certain cluster combinations may not be feasible, depending on the existing ward setup and capacity.

Finally the model is solved using CPLEX v12.5. The first numerical test with a data set from a large German hospital show that up to 3.5 % of required bed capacity can be saved by clustering departments and wards considering all relevant constraints, while maintaining the same service level. This amounts to a significant saving potential for the hospital. Furthermore, we compared our approach regarding bed requirement calculations with the Erlang-Loss approach and found our solution to be significantly more robust over multiple sets of simulated data. The model has been solved for various test instances within seconds.

Discussion

Bed capacity planning can have significant impact on perceived service and actual costs of a hospital. Optimal clustering of departments and wards will allow achieving balancing effects while keeping the increase of complexity reasonably low. We discuss the conceptual background, requirements and impact of such a clustering approach. Our literature review shows that an efficient modeling that also takes into account seasonality effects and constraints personnel management and qualification has not yet been developed. Our decision model is based on a mixed integer formulation using precalculated cost parameters. First computational tests show significant improvements for bed requirements and costs at a large German hospital.

References

de Bruin, A.M., Bekker, R., van Zanten, L., Koole, G.M.: Dimensioning hospital wards using the Erlang loss model. Ann. Oper. Res. **178**(1), 23–43 (2010)

Fügener, A., Hans, E.W., Kolisch, R., Kortbeek, N., Vanberkel, P.T.: Master surgery scheduling with consideration of multiple downstream units. Eur. J. Oper. Res. **239**(1), 227–236 (2014)

Gorunescu, F., McClean, S.I., Millard, P.H.: Using a queueing model to help plan bed allocation in a department of geriatric medicine. Health Care Manag. Sci. **5**(4), 307–312 (2002)

Lapierre, S.D., Goldsman, D., Cochran, R., DuBow, J.: Bed allocation techniques based on census data. Socioecon. Plann. Sci. **33**(1), 25–38 (1999)

Utley, M., Gallivan, S., Treasure, T., Valencia, O.: Analytical methods for calculating the capacity required to operate an effective booked admissions policy for elective inpatient services. Health Care Manag. Sci. **6**(2), 97–104 (2003)

van Essen, T.J., van Houdenhoven, M., Hurink, J.L.: Clustering clinical departments for wards to achieve a prespecified blocking probability. OR Spectr. **37**(1), 243–271 (2015)

Management of Blood Donation System: Literature Review and Research Perspectives

Seda Baş, Giuliana Carello, Ettore Lanzarone, Zeynep Ocak, and Semih Yalçındağ

Introduction

Applying optimization methods to healthcare management and logistics is a developing research area with numerous studies. Specifically, facility location, staff rostering, patient allocation, and medical supply transportation are the main themes analysed. Optimization approaches have been developed for several healthcare related problems, ranging from the resource management in hospitals to the delivery of care services in a territory. However, optimization approaches can also improve other services in the health system that have been only marginally addressed, yet. One of them is the Blood Donation (BD) system, aiming at providing an adequate supply of blood to Transfusion Centres (TCs) and hospitals.

Blood is necessary for several treatments and surgeries, and still a limited resource. The need for blood is about ten million units per year in the USA, 2.1 in Italy and 2 in Turkey; moreover, people still die in some countries because of inadequate supply of blood products (World Health Organization 2014). Hence, BD plays a fundamental role in healthcare systems, aiming at guaranteeing an adequate blood availability to meet the demand and save lives. In Western countries, blood is

S. Baş (✉) • Z. Ocak • S. Yalçındağ
Industrial and Systems Engineering Department, Yeditepe University, Istanbul, Turkey
e-mail: seda.bas@yeditepe.edu.tr; zocak@yeditepe.edu.tr; semih.yalcindag@yeditepe.edu.tr

G. Carello
Dipartimento di Elettronica, Informazione e Bioingegneria, Politecnico di Milano, Milan, Italy
e-mail: giuliana.carello@polimi.it

E. Lanzarone
Istituto di Matematica Applicata e Tecnologie Informatiche (IMATI), Consiglio Nazionale delle Ricerche (CNR), Milan, Italy
e-mail: ettore.lanzarone@cnr.it

A. Matta et al. (eds.), *Health Care Systems Engineering for Scientists and Practitioners*, Springer Proceedings in Mathematics & Statistics 169,
DOI 10.1007/978-3-319-35132-2_12

usually collected from *donors*, i.e., unpaid individuals who give blood voluntarily. Blood is classified into groups (A and subgroups, B, 0 or AB) and based on the Rhesus factor (Rh+ or Rh-), and each donor should be correctly matched with the patient who receives his/her blood. Moreover, as it may transmit diseases, blood must be screened before utilization.

Generally, there are two types of donation: whole-blood donation, in which the whole blood is directly collected in a plastic bag, and *apheresis*, i.e., the donation of specific components in which a mechanical gathering unit decays the required blood parts.

Blood requires particular precautions for collection and storage, and its shelf life from donation to utilization is limited, thus requiring a continuous feeding of the system (Greening et al. 2010). Hence, a successful BD supply chain should meet the daily demand of blood and follow its temporal pattern. According to Sundaram and Santhanam (2011), BD supply chain and the related management problems can be classified based on the main phases of a blood bag life: donor registration, blood collection, blood screening/evaluation, inventory storage and delivery. A slightly different classification is proposed in Pierskalla (2004), according to which the management of BD supply chain concerns both strategic decisions (e.g., location of blood centres) and tactical operational decisions (e.g., production of multiple products, control of inventory levels, blood allocation to hospitals, and delivery to multiple sites). In our review, we refer to the first classification scheme.

Many papers address the management of the BD supply chain (see Belien and Forcé (2012) for a recent survey); however, there are still some open issues. The aim of this paper is reviewing the literature related to the BD system management and classifying the existing research based on the process phase, in order to highlight unexplored issues and to point out alternative perspectives and possible future research opportunities. In section "Phases of Blood Donation System" we give details about the BD system and survey the existing literature (review updated at December 2014); then in section "Discussion and Open Issues" we discuss the open issues and propose future research directions.

Phases of Blood Donation System

BD supply chain can be divided into four main steps, as reported in Fig. 1: collection, transportation, storage and utilization. First, the blood is collected: donors are checked in blood centres to assess their eligibility and, if eligible, they make the donation. Once the blood is gathered, tests are independently performed on each individual's blood in order to prevent infectious diseases (screening process). Afterwards, the blood is transported and stored. Components are then distributed to the hospitals based on their inventory needs. Finally, it is transferred to the final users for transfusion.

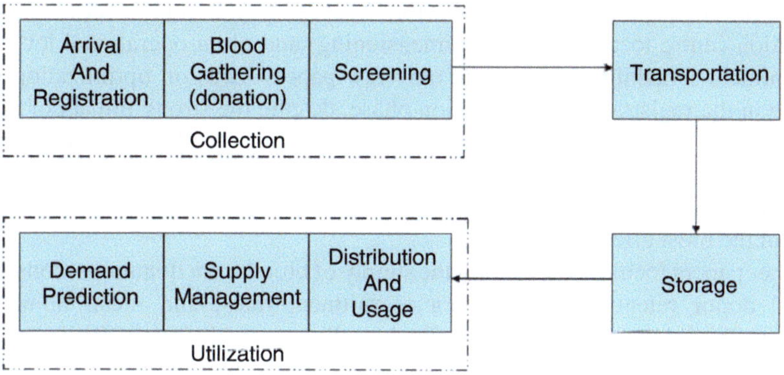

Fig. 1 Phases of the life cycle of a blood bag

Donors, Blood Collection and Screening

BD process starts with the arrival of the donor at the blood center. Donors can be divided in returning donors, who donate on an almost regular basis, and walk-in donors, who are entering the system occasionally or for the first time. In any case, donations can be made after a defined rest period from the previous one, which is defined by law. As donors have a crucial importance in the system, their availability, frequency and motivation have been studied from both a statistical and a social perspective.

Social Aspects The main reasons for blood donation and their relative importance have been studied by Bani and Giussani (2011). Moreover, it is also documented that the organization of blood collection phase may have an impact on donors' availability. Poor treatment, poor staff skills, and a bad experience are the main reasons of not returning to donate (Schereiber et al. 2003). Also prolonged queuing times are negatively correlated to BD satisfaction (McKeever et al. 2006; Katz et al. 2007). Hence, a well-organized donation management has a strong impact on the availability of blood bags, and also on donors' motivation, thus possibly increasing/decreasing their availability.

Donor Arrival and Registration When a donor enters in the system for the first time, he/she is requested to provide personal (e.g., name, address, age, job, gender) and medical/health (e.g., diagnosis, lab results, treatments) data, which are digitally collected. Digital registration provides a good traceability of the transfusion cycle, from collection to blood distribution and transfusion. The registration also includes a visit from a physician, followed by blood exams. If the donor is eligible, blood collection centres check that he/she makes the first donation within few days from the declaration of eligibility. Sometimes, the first visit is directly followed by a donation. A visit is also made before each donation or exam, during which the donor is re-evaluated and his/her personal data are updated.

Several management problems arise, both at a planning level (e.g., blood collection centre location or staff dimensioning) and at an operational level (e.g., appointment scheduling). However, only few papers focus on optimization issues arising in the registration and donation phase, despite the strong impact of donors' arrivals on the overall system performance. Michaels et al. (1993) developed a simulation study to evaluate scheduling strategies of donors arriving at a Red Cross blood drive, and compared these strategies in terms of donors' mean transit time to find out the most effective one.

Other papers focus on estimating the supply of blood from donations, considering annual donor retention rates, donor recruitment rates, and mean numbers of donations per donor and per year (Borkent-Raven et al. 2010). Ritika and Pau (2014) examined different classification algorithms to find out a fair classification technique for the prediction of donations. Flegel et al. (2000) developed a logistic regression model to compute the donation probability within a given time frame, considering different regression coefficients for walk-in and returning donors. Ferguson and Bibby (2002) used a prospective design to predict the number of future blood donations. Boonyanusith and Jittamai (2012) investigated the pattern of donors' behaviors based on factors influencing blood donation decision using a questionnaire.

Finally, on-line applications and database systems for donors' and bags management are also investigated (Chau et al. 2010; Khan and Qureshi 2009; Kulshreshtha and Maheshwari 2011).

Blood Collection and Screening Blood collection centres should be located according to their accessibility from hospitals in order to improve the overall system performance. Moreover, centres are generally subject to regulatory control, designed to ensure the maximum quality and safety of blood products. They guarantee that blood bags are produced according to standardized procedures, to achieve consistency of each product (Council of Europe 2007).

Despite the importance of this phase, the literature on blood collection system planning is rare (Ofori et al. 2005; WHO 2008; Lieshout-Krikke et al. 2013). De Angelis et al. (2003) proposed a stochastic methodology to analyse and certify the optimal configuration of servers by integrating simulation and optimization for a transfusion centre in Rome.

After collection, the screening phase starts with few tests performed against infectious diseases, e.g., HIV, Hepatitis B and C, and syphilis. They are repeated on each gathered blood sample, and are generally the same all around the world. Based on screening results, the blood bag is either released for clinical and manufacturing use or discarded (WHO 2008, 2010). An effective, well-organized screening program and a good quality system are essential for provisioning safe blood bags to patients and meeting the transfusion requirements.

Transportation and Storage of Blood Products

Once collected from donors at regional or community blood centres, blood must be stored in storage centres or TC before it perishes. These locations serve as a depot until the blood is distributed to the demand points and sometimes deal with testing of the blood products.

If collection and storage or TC centres do not coincide, blood must be transported. Although transportation is a rather simple task in this phase because all collected bags are usually addressed to a big TC or storage centre from all BD centres in the related territory, transportation must be carefully performed as the blood must be stored before perishing and requires particular transportation conditions. Inefficient and inadequate transportation may reduce the quality of end user care and increase costs. There is not much literature available about blood transportation between collection and storage centres; on the contrary, many papers on blood transportation focus on the distribution to hospitals (see subsection "Distribution and Utilization"). Ghandforoush and Sen (2010) used a deterministic non-convex integer optimization model to schedule the shuttle transportation of whole blood products from the collection points to the regional processing centres.

The presence of blood collection vehicles is also considered in blood bag transportation. Ekici and Ozener (2014) defined a variant of the Vehicle Routing Problem, i.e., the Maximum Blood Collection Problem (MBCP), in which blood collected in a set of blood donation sites is delivered with a fleet of collection vehicles to a single processing centre. Usually, there is no capacity limitation on the vehicles due to small size of the blood collection bags (Yi 2003; Doerner et al. 2008). On the contrary, time constraints are important because donated blood has to be delivered to the processing centre within a certain amount of time.

More attention has been paid to the storage of blood products. During the past 20 years significant progresses have been made in the technology of blood component preparation and storage (McCullough 2005; Blajchman et al. 1979). Belien and Forcé (2012) included several works in their survey.

Literature is mainly focused on inventory management problems (Padmanabhan and Vrat 1995; Axsäter 1996; Giri and Chaudhuri 1998; Dye and Ouyang 2005; Parlar et al. 2011), from both a deterministic (Padmanabhan and Vrat 1995; Axsäter 1996; Giri and Chaudhuri 1998; Jayaraman et al. 2010; Lieshout-Krikke et al. 2013) and a stochastic perspective (Prastacos 1978; Sirelson and Brodheim 1991; De Angelis et al. 2003; Pereira 2005; Katsaliaki 2008; Kopach et al. 2008; Blake 2009; Van Dijk et al. 2009; Parlar et al. 2011; Alfonso et al. 2012). Sirelson and Brodheim (1991) built a stochastic simulation model as a function of base stock levels to manage inventory level, outdated performance measures and shortage rates. Pereira (2005) built a stochastic model for a hospital blood bank inventory system, in which the remaining shelf life of blood units and the number of days between consecutive shipments were analysed according to the daily transfusion mean and variation impact. Katsaliaki (2008) used a stochastic simulation model for a cost-effective management of blood in the UK: valuable recommendations are provided to the stakeholders for cost reductions and for increasing the level of services and safety of the processes. Pierskalla and Roach (1972) grouped stock levels into categories according to shelf age; to satisfy the current (deterministic) demand, a First-In First-Out (FIFO) optimal policy was then applied issuing the oldest unit. Kopach et al. (2008) revisited a queuing model and, using level crossing techniques, determined an optimal policy to support the modeling of several trade-offs; the model was also combined with the current control techniques using simulation and the effectiveness of the model was verified with real data. Hemmelmayr et al. (2009) evaluated the impact of switching from their present vendee (customer) managed inventory system to a vendor (supplier) managed inventory system via a stochastic integer programming-based approach.

Some researchers extended inventory models like the Economic Order Quantity (EOQ) policy for including perishable products. For example, Giri and Chaudhuri (1998) proposed an inventory model for a perishable product where the demand rate is a function of the on-hand inventory, and the holding cost is non-linear. Padmanabhan and Vrat (1995) proposed a stock-dependent selling rate model where the backlogging function was assumed to be dependent on the amount of demand backlogged. Dye and Ouyang (2005) extended their model by introducing a time-proportional backlogging rate.

Distribution and Utilization

The last step of the BD chain includes distribution and utilization, which involve several management problems as detailed below. Distribution is highly important for efficient blood usage and should meet the demand, which is often uncertain and requires accurate predictions.

Demand Prediction for Blood Products Several works include an evaluation of the demand, even if general papers that only focus on a stochastic prediction of the demand are not available. They can mainly classified based on the demand structure: deterministic (Pierskalla and Roach 1972; Prastacos and Brodheim 1980; Hirsch and Brodheim 1981; Hirsch and Cazal 1981; Sahin et al. 2007; Ghandforoush and Sen 2010) or stochastic (Kaspi and Perry 1983; Jagannathan and Sen 1991; Custer et al. 2005; Pereira 2005; Katsaliaki and Brailsford 2007; Sahin et al. 2007; Katsaliaki 2008; Kopach et al. 2008; Haijema et al. 2009; Van Dijk et al. 2009; Hemmelmayr et al. 2010; Delen et al. 2011). Moreover, some of the works can also be classified with respect to the aggregation level: single hospital (Novis et al. 2002; Pereira 2005; Katsaliaki 2008; Blake 2009; Haijema et al. 2009; Perera et al. 2009; Van Dijk et al. 2009; Delen et al. 2011) or regional level (Glynn et al. 2003; Bosnes et al. 2005; Carden and DelliFraine 2005; Denesiuk et al. 2006; Katsaliaki and Brailsford 2007; Sahin et al. 2007; Erickson et al. 2008; Katsaliaki 2008; Kopach et al. 2008; Davis et al. 2009; Van Dijk et al. 2009; Hemmelmayr et al. 2009, 2010; Ghandforoush and Sen 2010). Some peculiarities of the process are also considered. For example, Kaspi and Perry (1983) considered a system in which both arrival of blood products and demand are modelled via stochastic process as independent Poisson processes. Silva Filho et al. (2012, 2013) developed a demand forecasting tool to make decisions about the weekly demand required by hospitals, and improved the planning of the inventory balance process with a strategy oriented to the forecasting of the demand of blood components. Forecasting the monthly demand was also investigated in Pereira (2004) by univariate time-series methods. Lau et al. (2013) predicted the future blood demand of thalassemia major patients for the next 10 years for long-term management of blood supply.

Management Policies Some papers deal with the decision making support in BD supply chain management, and on how to maintain or increase the supply of blood products (Sahin et al. 2007; Davis et al. 2009; Haijema et al. 2009; Van Dijk et al. 2009; Ghandforoush and Sen 2010; Hemmelmayr et al. 2010; Delen et al. 2011). Sahin et al. (2007) established several deterministic mathematical models to solve the location problem of blood services. Hemmelmayr et al. (2010) used an integer programming model to generate low costs and robust delivery routes for the supply of blood products to hospitals from a blood bank, and showed the impact of the uncertain demand on the resulting routes. Haijema et al. (2009) combined stochastic dynamic programming and simulation for the inventory management problem; the first approach is used to obtain optimal solutions, whereas the latter to investigate various what-if questions.

Distribution to Users and Usage Distribution starts with the delivering of components to hospitals, where they are transfused into patients. TCs are usually responsible for the provisioning of blood products to hospitals, and the delivered quantities are limited by the shelf-life of blood products as well as by the holding capacity. Two types of blood distribution systems were outlined by Hirsch and Cazal (1981): the reactive type, where the inventory level of the hospital is managed with respect to demand, and the predictive type, where the demand is fixed on schedule.

Prastacos and Brodheim (1980) focused on a deterministic mathematical programming model, whose target is to streamline the distribution of the regional blood resources while viewing plan commitments. It is characterized by a centralized management of blood rather than an individual hospitals management, pre-scheduled deliveries, and a distribution system in which blood is rotated among the hospitals.

Generally, redistributing the blood among hospitals is equally important for preventing out dating. In the event that there is a pressing need of a particular blood type in a clinic, they may use the blood with the closest decay date from an alternative facility to avoid spoilage of blood units. Kendall and Lee (1980) focused on this redistribution problem: their model has distinctive goals, e.g., anticipation of blood shortages and overages in hospitals, minimization of the quantity of old units, and minimization of the working expenses.

Recently, Le et al. (2013) combined inventory and routing management into one model: they proposed a column generation-based heuristics to solve the problem, and showed significant savings when using their model. Shen et al. (2003) presented a joint location-inventory model for blood distribution system, with non-linear working-inventory costs and non-linear safety stock inventory costs.

Discussion and Open Issues

Our analysis points out the high number of papers related to the management of storage and distribution phases. Indeed, Fig. 2 shows the percentage of the existing works for each phase. It can be seen that, even though the arrival of donors and the registration and donation system strongly affect the entire BD chain, only the 1 % of the investigations are devoted to improve these aspects. Hence, we found out the necessity of more adequate analyses and studies for this phase.

In particular, a relevant problem is the management of donors' appointments and visits, as it has a significant impact on the effectiveness of the entire BD chain and on donors' motivation. Increasing the number of donations improves the performance of the system, but also an effective management of donors' arrivals along with the days may optimize the daily production of bags with respect to the demand. Indeed, an unbalanced feeding of blood bags undermines the entire BD chain; this is not only a theoretical problem, but from the discussion with several blood providers this is the actual bottleneck of the entire system in the practice. Returning donors' appointments could be scheduled in advance, but not all donors are willing to accept pre-scheduled appointments, or they often require appointments at the beginning or at the ending of the day rather than at noon. Thus, an important future research is the development of optimization models and techniques for providing an efficient appointment scheduling, also in the light of balancing the production. The existing studies solved these problems by using simulation models (see, e.g., Lailomthong and Prichanont 2014); however, they do not fit the DB system since they do not take walk-in donors into account. The historical data collected by the BD centres can be exploited in these models, to forecast the walk-in donors' arrivals and increase the efficiency of the system. An effective application system is also needed in BD, as

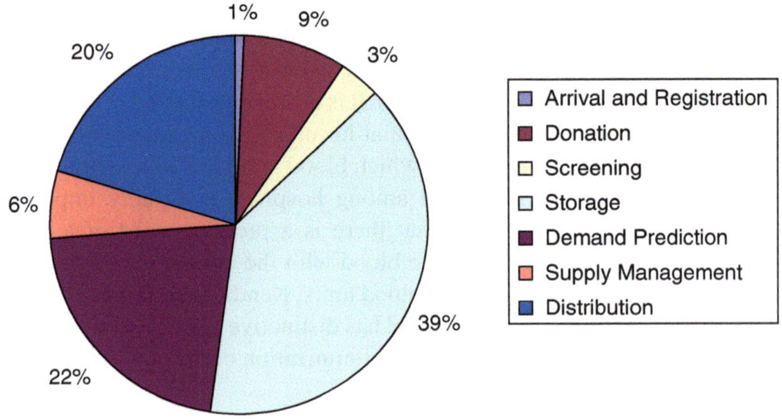

Fig. 2 Percentage of the existing works for each phase , considering 156 papers on blood management found in the literature (research updated at December 2014; papers on social and physiological aspect neglected)

in other domains, to combine the registration system with donors' and physicians' preferences and their points of view. Such an application system (e.g., an on-line system) could be a solution to join donors and physicians at the same platform and to encourage volunteer donations.

Storage is another important step of the system. A successful storage management should guarantee a proper balancing between the blood to hold and that to transfer, to keep blood in optimal conditions and to avoid expiring and discharging. This also stresses the importance of an adequate feeding with respect to the demand. The storage problem is widely studied in the literature (the 39 % of the investigations in Fig. 2). Existing models are generally based on the analysis of the normalized stock level, and they aim at predicting and reducing outdated bags and blood shortage. Nevertheless, an integrated management with blood feeding, i.e., with donor appointment scheduling, might increase the efficiency of the whole BD chain and reduce both outdated bags and blood shortage.

As mentioned, demand prediction is another crucial issue in BD system management. Inaccurate estimations of blood demand may lead to disruptive consequences. For example, underestimation leads to low quality of the service, out-of-stock and additional expenses; on the other hand, overestimation leads to overproduction and overstocking, together with increased costs and clinical and ethical problems in throwing bags away. Demand variation is an important factor to which the entire process must properly react; for example, blood inventory management becomes critical in case of increased demand, and the related decisions must be taken on time. However, meeting the demand is not easy since also the number of donors is difficult to foresee; hence, an integrated approach that considers the variation of both demand and donor arrivals should be required to better manage the BD chain.

Finally, transportation and delivery of blood products are largely addressed by means of optimization tools. Generally, the existing works deal with the routing of delivery vehicles for the distribution of blood components. As a future research line, with the increase in the use of blood components, an emerging logistics problem is the distribution of different products, while taking into account both their different shelf lives and cost minimization (multi criteria objective).

References

Alfonso, E., Xie, X., Augusto, V., Garraud, O.: Modeling and simulation of blood collection systems. Health Care Manag. Sci. **15**(1), 63–78 (2012)

Axsäter, S.: Using the deterministic EOQ formula in stochastic inventory control. Manag. Sci. **42**(6), 830–834 (1996)

Bani, M., Giussani, B.: Motivation in Italian whole blood donors and the role of commitment. Psychol. Health Med. **16**(6), 641–649 (2011)

Belien, J., Forcé, H.: Supply chain management of blood products: a literature review. Eur. J. Oper. Res. **217**(1), 1–16 (2012)

Blajchman, M.A., Shepherd, F.A., Perrault, R.A.: Clinical use of blood, blood components and blood products. Can. Med. Assoc. J. **121**(1), 33–42 (1979)

Blake, J.T.: On the use of operational research for managing platelet inventory and ordering. Transfusion **49**(3), 396–401 (2009)

Boonyanusith, W., Jittamai, P.: Blood donor classification using neural network and decision tree techniques. In: Proceedings of the World Congress on Engineering and Computer Science WCECS, pp. 499–503 (2012)

Borkent-Raven, B., Janssen, M., Poel, C.: Demographic changes and predicting blood supply and demand in the Netherlands. Transfusion **50**(11), 2455–2460 (2010)

Bosnes, V., Aldrin, M., Heier, H.E.: Predicting blood donor arrival. Transfusion **45**(2), 162–170 (2005)

Carden, R., DelliFraine, J.L.: An examination of blood center structure and hospital customer satisfaction: What can centralized and decentralized blood centers learn from each other? Health Mark. Q. **22**(3), 21–42 (2005)

Chau, M., Cheng, E. Chan, C.W.: Data analysis for healthcare: a case study in blood donation center analysis. In: Proceedings of 16th Americas Conference on Information Systems AMCIS, p. 242 (2010)

Council of Europe: Guide to the Preparation, Use and Quality Assurance of Blood Components, ISBN: 9789287161376 (2007)

Custer, B., Johnson, E.S., Sullivan, S.D., et al.: Community blood supply model: development of a new model to assess the safety, sufficiency, and cost of the blood supply. Med. Decis. Mak. **25**(5), 571–582 (2005)

Davis, R., Geiger, B., Gutierrez, A., et al.: Tracking blood products in blood centres using radio frequency identification: a comprehensive assessment. Vox Sang. **97**(1), 50–60 (2009)

De Angelis, V., Felici, G., Impelluso, P.: Integrating simulation optimisation in health care centre management. Eur. J. Oper. Res. **150**(1), 101–114 (2003)

Delen, D., Erraguntla, M., Mayer, R.J., Wu, C.N.: Better management of blood supply chain with GIS based analytics. Ann. Oper. Res. **185**(1), 181–193 (2011)

Denesiuk, L., Richardson, T., Nahirniak, S., Clarke, G.: Implementation of a redistribution system for near-outdate red blood cell units. Arch. Pathol. Lab. Med. **130**(8), 1178–1183 (2006)

Doerner, K.F., Gronalt, M., Hartl, R.F., et al.: Exact and heuristic algorithms for the vehicle routing problem with multiple interdependent time windows. Comput. Oper. Res. **35**(9), 3034–3048 (2008)

Dye, C., Ouyang, L.: An eoq model for perishable items under stock-dependent selling rate and time-dependent partial backlogging. Eur. J. Oper. Res. **163**(3), 776–783 (2005)

Ekici, A., Ozener, O.O.: Routing for blood supply management. In: 20th Conference of the International Federation of Operational Research Societies IFORS (2014)

Erickson, M.L., Champion, M.H., Klein, R., Ross, R.L., et al.: Management of blood shortages in a tertiary care academic medical center: the Yale-New Haven Hospital frozen blood reserve. Transfusion **48**(10), 2252–2263 (2008)

Ferguson, E., Bibby, P.A.: Predicting future blood donor returns: past behavior, intentions, and observer effects. Health Psychol. **21**(5), 513–518 (2002)

Flegel, W.A., Besenfelder, W., Wagner, F.F.: Predicting a donor's likelihood of donating within a preselected time interval. Transfus. Med. **10**(3), 181–192 (2000)

Ghandforoush, P., Sen, T.K.: A DSS to manage platelet production supply chain for regional blood centers. Decis. Support Syst. **50**(1), 32–42 (2010)

Giri, B., Chaudhuri, K.: Deterministic models of perishable inventory with stock-dependent demand rate and nonlinear holding cost. Eur. J. Oper. Res. **105**(3), 467–474 (1998)

Glynn, S.A., Busch, M.P., Schreiber, G.B., et al.: Effect of a national disaster on blood supply and safety: the September 11 experience. J. Am. Med. Assoc. 289(17), 2246–2253 (2003)

Greening, D.W., Glenister, K.M., Sparrow, R.L., Simpson, R.J.: International blood collection and storage: clinical use of blood products. J. Proteom. **73**(3), 386–395 (2010)

Haijema, R., van Dijk, N., van der Wal, J., Sibinga, C.S.: Blood platelet production with breaks: optimization by SDP and simulation. Int. J. Prod. Econ. **121**(2), 464–473 (2009)

Hemmelmayr, V., Doerner, K.F., Rfand, H., Savelsbergh, M.W.P.: Delivery strategies for blood products supplies. OR Spectr. **31**(4), 707–725 (2009)

Hemmelmayr, V., Doerner, K.F., Hartl, R.F., Savelsbergh, W.P.: Vendor managed inventory for environments with stochastic product usage. Eur. J. Oper. Res. **202**(3), 686–695 (2010)
Hirsch, R.L., Brodheim, E.: Blood distribution systems and the exchange of information between hospital blood banks and regional blood centers. Vox Sang. **40**(3), 239–244 (1981)
Hirsch, R.L., Cazal, P.: Recommendations of the task force on record-keeping and blood distribution systems. Vox Sang. **40**(3), 216–221 (1981)
Jagannathan, R., Sen, T.: Storing cross matched blood: a perishable inventory model with prior allocation. Manag. Sci. **37**(3), 251–266 (1991)
Jayaraman, S., Chalabi, Z., Perel, P., et al.: The risk of transfusion-transmitted infections in sub-Saharan Africa. Transfusion **50**(2), 433–442 (2010)
Kaspi, H., Perry, D.: Inventory systems of perishable commodities. Adv. Appl. Probab. **15**(3), 674–685 (1983)
Katsaliaki, K.: Cost-effective practices in the blood service sector. Health Policy **86**(2), 276–287 (2008)
Katsaliaki, K., Brailsford, S.C.: Using simulation to improve the blood supply chain. J. Oper. Res. Soc. **58**(2), 219–227 (2007)
Katz, L.M., Cumming, P.D., Wallace, E.L.: Computer-based blood donor screening: a status report. Transfus. Med. Rev. **21**(1), 13–25 (2007)
Kendall, K.E., Lee, S.M.: Formulating blood rotation policies with multiple objectives. Manag. Sci. **26**(11), 1145–1157 (1980)
Khan, A.R., Qureshi, M.S.: Web-based information system for blood donation. Int. J. Digit. Content Technol. Appl. **3**(2), 137–142 (2009)
Kopach, R., Balcioglu, B., Carter, M.: Tutorial on constructing a red blood cell inventory management system with two demand rates. Eur. J. Oper. Res. **185**(3), 1051–1059 (2008)
Kulshreshtha, V., Maheshwari, S.: Blood bank management information system in India. Int. J. Eng. Res. Appl. **1**(2), 260–263 (2011)
Lailomthong, N., Prichanont, S.: Patient's waiting time reduction in outpatient department. In: International Conference on Advances in Engineering and Technology ICAES, pp. 468–473 (2014)
Lau, E., He, X., Lee, C., Wu, J.: Predicting future blood demand from thalassemia major patients in Hong Kong. PLoS One **8**(12), e81846 (2013)
Le, T., Diabat, A., Richard, J. P., Yih, Y.: A column generation-based heuristic algorithm for an inventory routing problem with perishable goods. Optim. Lett. **7**(7), 1481–1502 (2013)
Lieshout-Krikke, R.W., Zaaijer, H.L., Prinsze, F.J.: The yield of temporary exclusion of blood donors, exposed to emerging infections abroad. Vox Sang. **104**(1), 12–18 (2013)
McCullough, J.: Transfusion Medicine, 2nd edn. Churchill Livingstone, London (2005), http://eu.wiley.com/WileyCDA/WileyTitle/productCd-144433705X.html
McKeever, T., Sweeney, M.R., Staines, A.: An investigation of the impact of prolonged waiting times on blood donors in Ireland. Vox Sang. **90**(2), 113–118 (2006)
Michaels, J.D., Brennan, J.E., Golden, B.L.: A simulation study of donor scheduling systems for the American Red Cross. Comput. Oper. Res. **20**(2), 199–213 (1993)
Novis, D.A., Renner, S., Friedman, R., et al.: Quality indicators of blood utilization: three college of American pathologists Q-probes studies of 12,288,404 red cell units in 1639 hospitals. Arch. Pathol. Lab. Med. **126**(2), 150–156 (2002)
Ofori, S.O., Temple, J., Sarkodie, F.: Predonation screening of blood donors with rapid test. Transfusion **45**(2), 133–140 (2005)
Padmanabhan, G., Vrat, P.: Eoq models for perishable items under stock dependent selling rate. Eur. J. Oper. Res. **86**(2–3), 281–292 (1995)
Parlar, M., Perry, D., Stadje, W.: FIFO versus LIFO issuing policies for stochastic perishable inventory systems. Methodol. Comput. Appl. Probab. **13**(2), 405–417 (2011)
Pereira, A.: Performance of time-series methods in forecasting the demand for red blood cells transfusion. Transfusion **44**(5), 739–746 (2004)
Pereira, A.: Blood inventory management in the type and screen era. Vox Sang. **89**(4), 245–250 (2005)

Perera, G., Hyam, C., Taylor, C., Chapman, J.F.: Hospital blood inventory practice: the factors affecting stock level and wastage. Transfus. Med. **19**(2), 99–104 (2009)

Pierskalla, W.P.: Supply chain management of blood banks (Chapt. 5). In: Brandeau, M.L., Sainfort, F., Pierskalla, W.P. (eds.) Operations Research and Health Care. Springer International Series in Operations Research & Management Science. Springer, New York, pp. 103–145 (2004)

Pierskalla, W.P., Roach, C.D.: Optimal issuing policies for perishable inventory. Manag. Sci. **18**(11), 603–614 (1972)

Prastacos, G.P.: Optimal myopic allocation of a product with fixed lifetime. J. Oper. Res. Soc. **29**(9), 905–913 (1978)

Prastacos, G.P., Brodheim, E.: PBDS: a decision support system for regional blood management. Manag. Sci. **26**(5), 451–463 (1980)

Ritika, Pau, A.: Prediction of blood donors population using data mining classification technique. Int. J. Adv. Res. Comput. Sci. Softw. **4**(6), 634–638 (2014)

Sahin, G., Sural, H., Meral, S.: Locational analysis for regionalization of Turkish Red Crescent blood services. Comput. Oper. Res. **34**(3), 692–704 (2007)

Schereiber, G.B., Glynn, S.A., Damesyn, M.A., et al.: Lapsed donors: an untapped resource. Transfusion **43**(1), 17–24 (2003)

Shen, Z., Coullard, C., Daskin, M.: A joint location-inventory model. Transp. Sci. **37**(1), 40–55 (2003)

Silva Filho, O.S., Cezarino, W., Salviano, G.R.: A Decision-making tool for demand forecasting of blood components. In: 14th IFAC Symposium on Information Control Problems in Manufacturing, pp. 1499–1504 (2012)

Silva Filho, O.S., Carvalho, M.A., Cezarino, W., et al.: Demand forecasting for blood components distribution of a blood supply chain. In: 6th IFAC Conference on Management and Control of Production and Logistics, pp. 565–571 (2013)

Sirelson, V., Brodheim, E.: A computer planning model for blood platelet production and distribution. Comput. Methods Programs Biomed. **35**(4), 279–291 (1991)

Sundaram, S., Santhanam, T.: Comparison of blood donor classification data mining models. J. Theor. Appl. Inf. Technol. **30**(2), 98–101 (2011)

Van Dijk, N., Haijema, R., van der Wal, J., Sibinga, C.S.: Blood platelet production: a novel approach for practical optimization. Transfusion 49(3), 411–420 (2009)

WHO: Global Database on Blood Safety, 2004–2005 Report. World Health Organization, Geneva (2008)

WHO: Screening Donated Blood for Transfusion Transmissible Infections - Recommendations: http://www.who.int/bloodsafety/ScreeningDonatedBloodforTransfusion.pdf (2010)

World Health Organization: http://www.who.int (2014)

Yi, J.: Vehicle routing with time windows and time-dependent rewards: a problem from the American Red Cross. Manuf. Serv. Oper. Manag. **5**(1), 74–77 (2003)

Staffing Ratio Analysis in Primary Care Redesign: A Simulation Approach

Xiang Zhong, Hyo Kyung Lee, Molly Williams, Sally Kraft, Jeffery Sleeth, Richard Welnick, Lori Hoschild, and Jingshan Li

Introduction

Primary care physicians are facing an enormous amount of tasks which make them at high risk of burnout. However, many functions they are performing do not require their professional training (Sinsky et al. 2010, 2013). To alleviate this situation, team-based care, in which support staff work collaboratively with patients and their caregivers to achieve coordinated and high-quality care, establishes a cornerstone of the modern primary care model (Patient Centered Primary Care Implementation Work Group; Doherty & Crowley 2013; Patel et al. 2013a). These redesigning efforts share a common direction: more emphasis on expanding and elevating the role of team members with lower level clinical credentials (Berra). Expanding staffing and shifting roles offers possible advantages. However, the question of what should be the right staffing ratio between support staff and physician is still unanswered. In recent years, such a question has been studied prevalently (Berra; Wharrad & Robinson 1999; OECD 2007; Reinhardt et al. 1972; Peikes et al. 2014; Patel et al. 2013b; Reid et al. 2010). Multiple ratios, such as 2.6:1 (Berra), 2:1 (Wharrad & Robinson 1999), 2.9–3.1:1 (OECD 2007), 4.25:1 (Patel et al. 2013b), have been proposed for various systems. Thus, the optimal staffing ratio is still unclear. This motivates our study to identify whether there is a fundamental rule governing optimal staffing ratio under various practice settings.

X. Zhong (✉) • H.K. Lee • J. Li
University of Wisconsin-Madison, 1513 University Ave., Madison, WI 53705, USA
e-mail: xzhong4@wisc.edu

M. Williams • J. Sleeth • R. Welnick • L. Hoschild
University of Wisconsin Medical Foundation, Middleton, WI 53562, USA

S. Kraft
Dartmouth-Hitchcock, Hanover, NH 03755, USA

A. Matta et al. (eds.), *Health Care Systems Engineering for Scientists and Practitioners*, Springer Proceedings in Mathematics & Statistics 169,
DOI 10.1007/978-3-319-35132-2_13

The remainder of the paper is structured as follows: Section "Literature Review" introduces the method of the study. The results are reported in section "Problem Addressed". Finally, section "Model Description" summarizes the takeaways and outlines the future work.

Methods

Primary Care Patient Flow

A general primary care visit can be described as follows: A patient arrives at the primary care site, greeted by a receptionist. After check-in, an MA will escort the patient to an exam room. The MA completes vitals, helps the patient identify the goals of visit, and completes the rooming workflow. In addition, the MA may administer any needed injections or immunizations prior to the physician visit (this might require a RN in some practices). Then the physician meets the patient, reconciles medications and updates patient's problem list. The physician may order medications, assess the patient's educational needs, and develop a plan of care with the patient. If education is needed, the proper educational/self-management tools for patient will be provided by the MA, RN or MD (as appropriate). After the physician visit, the patient will be given an after visit summary (AVS), and the MA may review it with the patient. Scheduling the patient for any primary care follow-up or specialty appointments will be completed by the MA (or receptionist in some practices). Finally, the patient leaves the facility when all services are complete. A general process flow of the primary care patient visit is illustrated in Fig. 1.

Data Collection

Two case studies were conducted at one pediatric and one adult primary care clinic owned and operated by UW Health in Madison, Wisconsin. These studies provide us

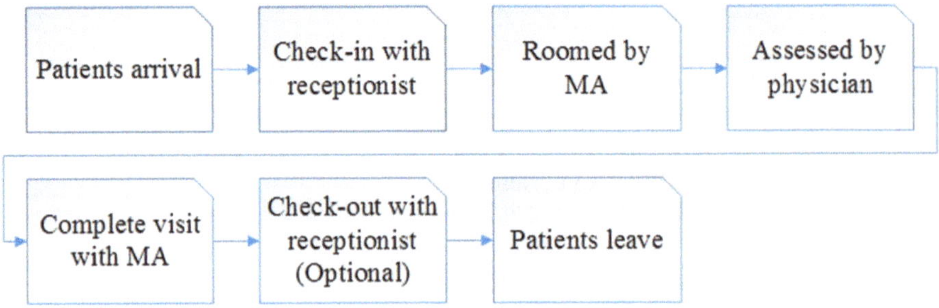

Fig. 1 Primary care patient flow

the opportunity to investigate the current workforce and their roles in the care team, understand the patient flow in the primary care clinic, and collect patient arrival and service time data. The information gathered is used to develop a simulation model to analyze patient flow, evaluate its design options, and propose recommendations for improvement.

Model Development

Given the patient flow introduced above, a simulation model is developed using SIMUL8. We design a clinic with four physicians and for each physician, four exam rooms are assigned. The number of MAs is increased from 1 to 14 in order to simulate the staff ratio (MA to MD) changing from 0.25 to 3.5.

Two types of visits are defined in this model: office visit (OFV) and long office visit (LOV), where LOV is scheduled for patients with more complex care needs. Arrivals are generated based on the scheduling template. The first patient is scheduled to arrive at 9:00 a.m. and the last patient is scheduled to arrive no later than 4:00 p.m. OFV patients and LOV patients are scheduled in an alternative manner using a single booking method in increments of 30 min. To incorporate the randomness in arrivals, the deviation of the actual arrival time from the appointment time is captured using field data, and is modeled as a triangular distribution added to the scheduled time. In addition, appointment vacancy and patient no-show rates are not negligible. To model this scenario, some patients are selected to leave the system without being served.

In the designed clinic, four physicians are on duty daily and share work with a pool of MAs. Each physician is assigned four exam rooms. Responsibilities of an MA include rooming, administration of injections, immunizations, etc. The MA and MD work with patients on a first come, first served basis. A receptionist is responsible for patient registration and check-out.

Operation times of each service are collected either by direct measurement or through estimation by clinic staff. The operation times associated with different clinic staffs are provided in Table 1. The numbers in the table are in time unit,

Table 1 Service time distribution settings

Parameters	Mean	Std. Dev.	Distribution	Description
Dist LOV rooming	21.4	10.7	LogNormal	MA rooming LOV patient
Dist OFV rooming	17.1	8.6	LogNormal	MA rooming OFV patient
Dist LOV physician visit	40	20	LogNormal	MD visit LOV patient
Dist OFV physician visit	20	10	LogNormal	MD visit OFV patient
Dist discharge	10.7	5.4	LogNormal	MA discharge patient
Dist check-out	1	1	LogNormal	Checkout with receptionist
Dist early late	(−50, 5.9, 30)		Triangular	Deviation of actual arrival time
Appointment vacancy	10 %			Probability of no show

representing the relative time distributed among different procedures. As shown in (Reynolds et al. 2010; Wang et al. 2014; Zhong et al. 2015), the average length of stays (ALOSs) are practically independent of distribution type, and are mainly determined by coefficient of variation (CV). Such a property is also observed in this study. Thus, log-normal distribution characterized by mean and standard deviation is used for modeling operation times.

Using the information discussed above, a complete simulation model is developed. Three modules are included in the model: pre-rooming service, which includes patient arrival and label assignment (characterizing patient identity); in-room service, starting with MA rooming, followed by the provider identifying patients' needs and finished by MA completing the patient visit; and post-room service, which is a quick optional check-out at registration.

Results

The simulation model was run for 50 trials, with 10 days per trial to gain the necessary confidence intervals. To verify the model, time stamps were carefully traced to check every step performed by the simulation model.

Workload Allocation Between MA and MD

In this subsection, the workload allocation between MA and MD is changed and for each workload allocation, different staffing models are compared. The ratio of workload allocation (MA to MD) is altered in the following pattern: for the cases where MA has more workload than MD, the ratio is changed from 1:1 to 2:1 in increments of 0.2. In the cases where MA has less workload than MD, the ratio is changed from 0.4:1 to 1:1 in increments of 0.2. The patient's average length of stay is measured. The results are shown in Fig. 2.

Figure 2 sheds light on the workload allocation ratio that yields the lowest length of stay in the system. For 1 MA, staffing ratio of MA to MD is 0.25:1. Represented by the first sub plot in the chart, the patient's average length of stay increases when more workload is assigned to the MA. So the time in system is the smallest when workload allocation is 0.4:1 (note that workload allocations do not go below 0.4 in the analysis). For two MAs, staffing ratio of MA to MD is 0.5:1, so the time in system is the shortest when workload allocation is between 0.4:1 and 0.6:1, as illustrated by the second subplot in Fig. 2. When there are three MAs, the nadir of the curve shifts to the point where workload ratio equals 0.8:1, matching the staffing ratio 0.75:1. Similar trend is identified for four MAs. Workload ratio of 1:1 pinpoints the watershed where either an increase or decrease of workload would lead to an increase in patient ALOS. As the staffing ratio increases, the rebound point shifts right correspondingly. This implies that ALOS is shortest when workload is balanced between MA and MD. In other words, staffing ratio should be equal to the workload ratio.

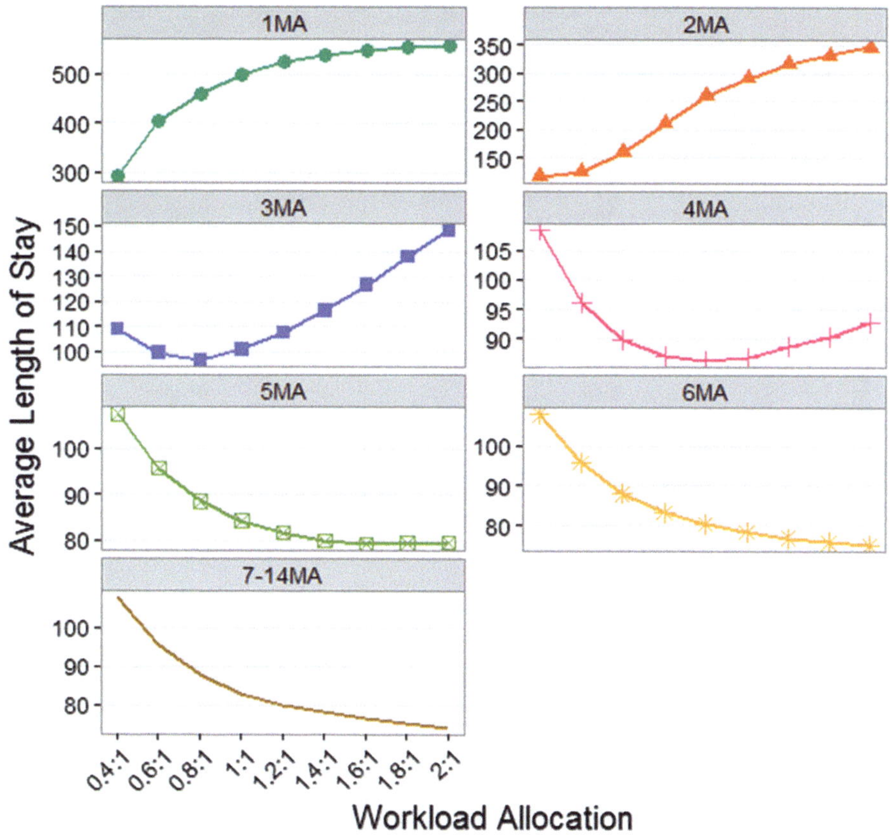

Fig. 2 Average length of stay of LOV patients

Remark 1: Throughout all sensitivity analyses, OFV patients' ALOS shares the same trend as LOV patients. Due to space limitation, the results of OFV patients are omitted.

Patient Volume

As patient volume varies daily and is highly affected by season, experiments under various patient populations are carried out. Compared to the original setting in which appointment vacancy was set to 10 %, probability of vacancy was decreased to 0 % to represent the busy days, and increased to 20 % to represent the inactive days. The comparison is shown in Figs. 3 and 4 below. As both figures show similar trends, the conjecture that time in system is shortest when workload allocation is balanced is invariant with patient volume.

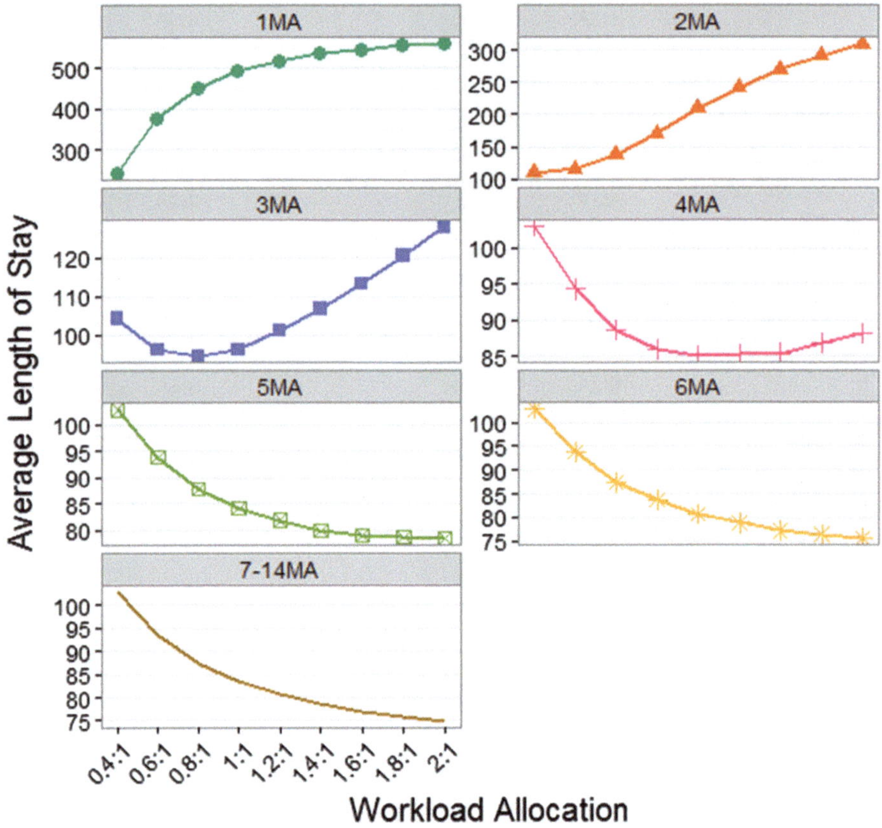

Fig. 3 Average length of stay of LOV under small patient volume

Variability in Service Times and Patient Arrival

Fundamental to the pursuit of better operations in healthcare delivery is understanding the fact that systems vary between and among patients, providers, and organizations. Such variations cannot be overlooked and how stochasticity affects the delivery of services and staffing model deserves a systematic study.

As a first illustration of variability, we allow the delivery of service to vary. Average service times remain the same but variation of service times is changed. We increase CV from 0.2 to 1.0 in increments of 0.2 and apply the same CV to all service distributions simultaneously at each CV level. The simulation results are summarized in Fig. 5. As one can see, service variations exert significant influence on average time in system. Variability causes congestion and delay, and thus patient's time in system is longer as variability of service is increased. Therefore, an effort to reduce variation of service can be beneficial.

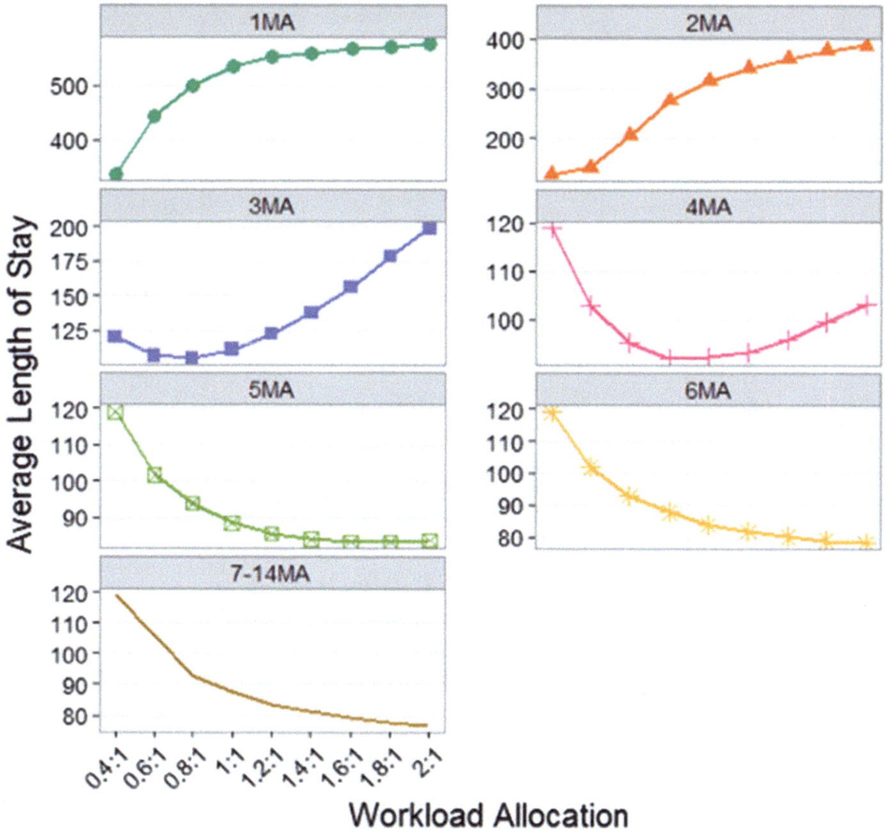

Fig. 4 Average length of stay of LOV under large patient volume

As for staffing levels, the changing pattern of ALOS with the increase of MA does not manifest significant difference among service time CVs. Regardless of CV change, ALOS decreases as the number of MA is increased up to a certain level. But the amount of difference diminishes as more MAs are added, implying that the effect of adding additional MAs weakens. This directs the focus of this section to find the optimal number of MA under various CV settings. The statistical tests show that there is significant decrease in ALOS when increased from 3 MAs to 4 MAs, while further increasing does not significantly decrease patient ALOS in the 95 % confidence level. This concludes that the performance comparison of staffing models is not significantly affected by service variations as long as they are in the same scale.

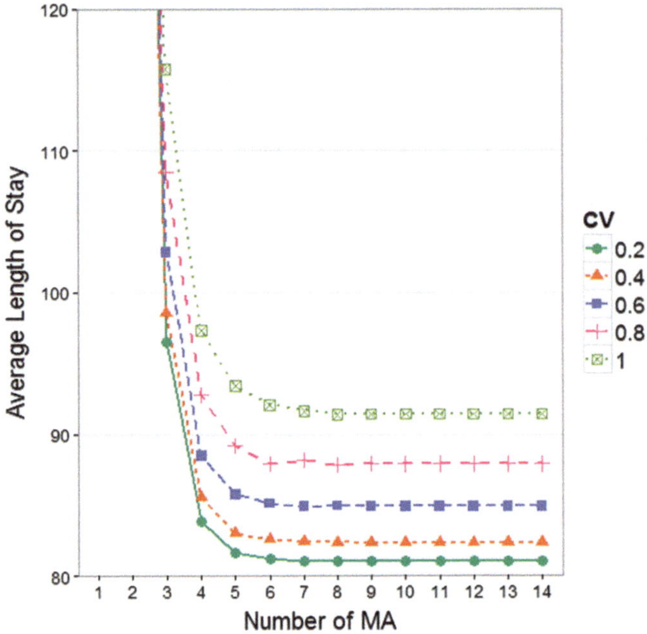

Fig. 5 Average length of stay of LOV patients under different service time distributions

In addition to service time variation, variability incurred by patient arrival is also considered. Here we change the variability of patient arrival by varying the deviation of patient arrival time from scheduled time. The lower bound and upper bound of the triangular distribution is adjusted to emulate small (−30, 5.92, 20), medium (−50, 5.92, 30) (original setting) and large variations (−90, 5.92, 60). Figure 6 shows the simulation results.

As shown in Fig. 6, time in system is longer as more variability is added to patient arrival, but the trend in which ALOS decreases with respect to the increase of MA is consistent regardless of arrival deviations. Similar to the variation in CV case, t-test was conducted to test the differences in ALOS. It shows that there is significant decrease as we increase from 3 MAs to 4 MAs, but further addition of MA has no significant effect on patient ALOS.

In this section, simulation results demonstrate how variations in service or patient arrival processes cause delays. System performance deteriorates when variation increases. However, performance comparison between different staffing models is not affected by the variability. If service times of MA and MD have the same scale of variation, the mean service time (equivalent to workload) is more critical in terms of deciding the optimal staffing model.

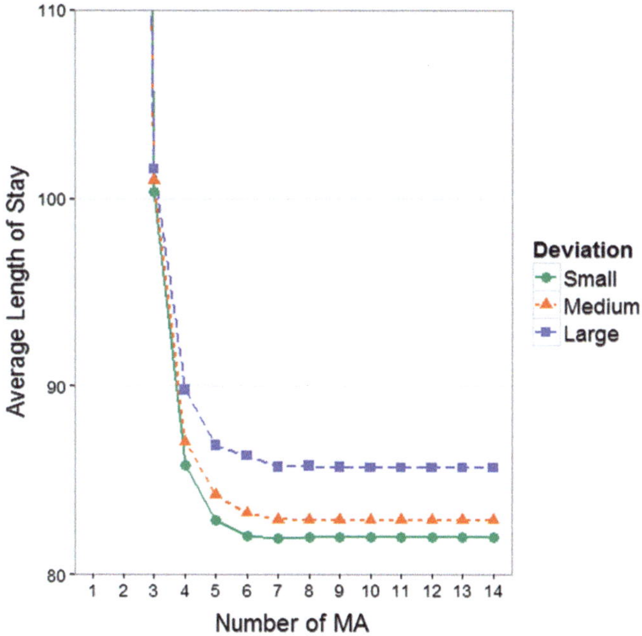

Fig. 6 Average Length of Stay of LOV patients under different arrival deviations

Number of Exam Rooms

In addition to staffing analysis, we also investigate the impact of other (non-staff) resources on system performance. Exam rooms are essential for providing care services. Variation in the number of exam rooms per MD is tested to see its potential impact on patient average length of stay.

Specifically, since five exam rooms per physician or one exam room is rarely observed in real practice, three scenarios: four exam rooms (original setting), three exam rooms, and two exam rooms per physician are considered. The resulting performances are compared in Fig. 7.

As apparent in Fig. 7, there is significant difference in patient ALOS between 2 and 3 rooms. But between 3 and 4 rooms, no significant difference is detected. However, increasing room number can hold more patients in the clinic, which increases patient throughput. Simulation results indicate that when room number is increased from 3 to 4, the number of completed patients increases by 12.24 %, about 20 patients per day. Thus, having four rooms per physician is a proper design in current clinic settings.

Remark 2: In addition to the pooled MA case, dedicated MA scenario is also investigated. The results show that the pooled MA performs better than the dedicated MA, so the ALOS achieved from the pooled MA model provides a lower bound of that of the dedicated MA model. Also, both average MA and MD utilization are

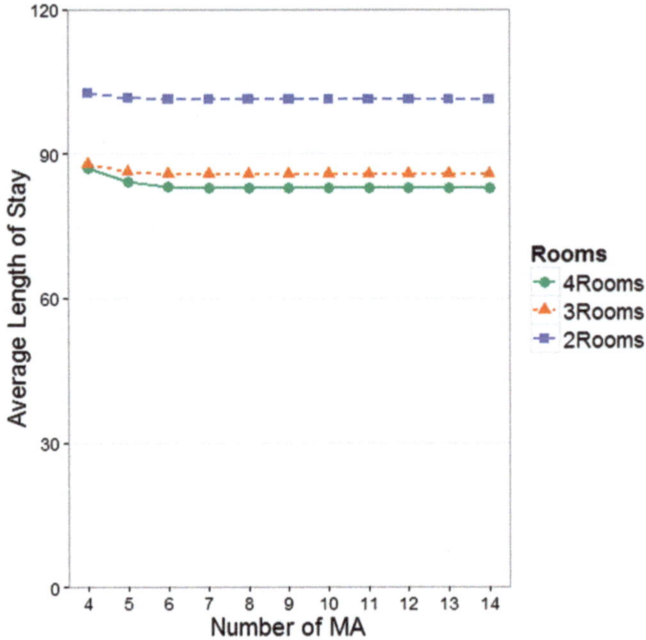

Fig. 7 Average length of stay of LOV patients under different number of exam rooms

higher when MAs are pooled. Thus even though a dedicated MA is largely used in practice, using pooled MA resources results in a more flexible and efficient system.

Conclusions

In close, primary care transformation represents a complex system redesign that requires a fundamental rule that aligns workload and staffing to support successful implementation of the redesign work.

Primary care practice should build care teams according to the workload of each team member. Specifically when assigning the work accommodating daily patient visits, total workload per physician and MA should be balanced. That is, the total workload (in contact with patient directly) of physicians (W_{MD}) divided by the number of physicians (N_{MD}) should be approximately equal to the total patient work of MAs (W_{MA}) divided by the number of MAs (N_{MA}):

$$\frac{W_{MD}}{N_{MD}} = \frac{W_{MA}}{N_{MA}}.$$

When the above equation is satisfied, patient average length of stay reaches the minimum and provider productivity is maximized.

If the staff composition is fixed, work can be designed to balance team member's workload. For instance, to alleviate the burden on physician, support staff such as medical assistant could take on expanded roles. As an example, the medical assistant's role has been transformed at North Shore Physician Group in the Boston area (Sinsky et al. 2013). The rooming process has been expanded from 3 to 8 min and includes medication review, agenda setting, form completion and closing care gaps. On the other hand, if the workloads are fixed, the team should try to hire the matching number of staff (FTE) to leverage. In the case of dedicated MA, the practice might alleviate the burnout of physicians and reduce waiting time of patients by adopting float MAs to share among providers.

The analyses described above could lead to a series of hypotheses for future research, including:

- Test the workload rule on different care models (not limited to primary care)
- Expand the model to incorporate RN, LPN and other team members. Investigate the total team composition to achieve best performance and utilization.
- Analyze delivery systems from a healthcare value perspective. Add cost model to assess the cost efficiency while also considering patient experience and staff satisfaction.

Acknowledgement This work is partially supported by NSF Grant No. CMMI-1233807.

References

Sinsky, C.A., Sinsky, T.A., Althaus, D., Tranel, J., Thiltgen, M.: 'Core teams': Nurse-physician partnerships provide patient-centered care at An Iowa practice. Health Aff. **29**(5), 966–968 (2010)

Sinsky, C.A., Willard-Grace, R., Schutzbank, A.M., Sinsky, T.A., Margolius, D., Bodenheimer, T.: In search of joy in practice: a report of 23 high-functioning primary care practices. Ann. Fam. Med. **11**(3), 272–278 (2013)

Patient Centered Primary Care Implementation Work Group, US Department of Veterans Affairs. Patient-aligned care team (PACT) (2012). http://www.va.gov/PrimaryCare/docs/pcmh_ConceptPaper.doc

Doherty, R.B., Crowley, R.A.: Health and Public Policy Committee of the American College of Physicians. Principles supporting dynamic clinical care teams: an American College of Physicians position paper. Ann. Intern. Med. **159**(9), 620–626 (2013)

Patel, M.S., Arron, M.J., Sinsky, T.A., Green, E.H., Baker, D.W., Bowen, J.L., Day, S.: Estimating the staffing infrastructure for a patient-centered medical home. Am. J. Manag. Care. **19**(6), 509–516 (2013a)

Berra, A. Benchmarking clinical support staff in primary care sites (2011). http://www.advisory.com/research/health-care-advisory-board/blogs/the-blueprint/2011/07/benchmarking-clinical-support-staff-in-primary-care-sites

Wharrad, H., Robinson, J.: The global distribution of physicians and nurses. J. Adv. Nursing **30**(1), 109–120 (1999)

Organisation for Economic Cooperation and Development (OECD): Health at a glance 2007: OECD indicators. Paris: OECD (2007)

Reinhardt, U.: A production function for physician services. Rev. Econ. Stat. **54**, 55–66 (1972)

Peikes, D.N., Robert, J.R., Timothy, J.D., Derekh, D.F.C., Stacy, B.D., Richard, J.B., Randall, S.B., Rachel, J.S.: Staffing patterns of primary care practices in the comprehensive primary care initiative. Ann. Fam. Med. **12**(2), 142–149 (2014)

Patel, M.S., Martin, J.A., Thomas, A.S., Eric, H.G., David, W.B., Judith, L.B., Susan, D.: Estimating the staffing infrastructure for a patient-centered medical home. Am. J. Manag. Care. **19**(6), 509–516 (2013b)

Reid, R.J., Coleman, K., Johnson, E.A., et al.: The group health medical home at year two: cost savings, higher patient satisfaction, and less burnout for providers. Health Aff. (Millwood) **29**(5), 835–843 (2010)

Wang, J., Zhong, X., Li, J., Howard, P.K.: Modeling and analysis of care delivery services within patient rooms: A system-theoretic approach. IEEE T. Autom. Sci. Eng. **11**(2), 379–393 (2014)

Reynolds, J., Zeng, Z., Li, J., Chiang, S.Y.: Design and analysis of a health care clinic for homeless people using simulations. Int. J. Health Care Qual. Assur. **23**(3), 607–620 (2010)

Zhong, X., Song, J., Li, J., Ertl, S.M., Fielder, L.: Analysis and design of Gastroenterology (GI) clinic in digestive health center: a systems approach. Flex. Serv. Manuf. (2015). doi:10.1007/s10696-015-9215-5

Disease Prevention and Control Plans: State of the Art and Future Research Guideline

Wanying Chen, Alain Guinet, and Angel Ruiz

Introduction

Outbreaks of large-scale diseases can pose a great threat to the public health and lead to severe consequences (WHO Global Influenza Preparedness Plan 2005). Now, the pollution, the urbanization and many other factors increase the possibility of the outbreaks of the diseases. In order to reduce or mitigate the threat and the consequences, an effective and comprehensive disease prevention and control plan is needed (WHO 2014). Now, more and more people study disease prevention and control plans. However, the research papers focusing on this field are still far from being enough. Moreover, to the best of our knowledge, until now, no paper has reviewed the articles about the disease prevention and control plans systematically. To fill this gap and shed light on future research directions, our paper studied the existing relevant articles according to different diseases, the related disease prevention and control plans and applied methodologies.

Basic Background

According to the infectiousness, diseases can be classified as communicable disease (CD) and non-communicable disease (NCD). NCD cannot be passed from person to person. CD can spread from one person to another or from an animal to a human

W. Chen (✉) • A. Guinet
DISP, INSA de Lyon, Bât. Jules Verne, 19 Av. Jean Capelle, 69621 Villeurbanne, France
e-mail: wanying.chen@insa-lyon.fr

A. Ruiz
CIRRELT, Université LAVAL, 2325 rue de la Terrasse, Québec, QC, Canada G1V 0A6

A. Matta et al. (eds.), *Health Care Systems Engineering for Scientists and Practitioners*, Springer Proceedings in Mathematics & Statistics 169,
DOI 10.1007/978-3-319-35132-2_14

being. In epidemiology, the basic reproduction number (R0), defined as the number of susceptible individuals that can be infected by each patient during the infectious periods, is always used to describe the infection rate of the communicable disease (Dasaklis et al. 2012). Disease prevention and control plans can be considered as strategies to guide the decision makers to adopt the appropriate disease response policy and accelerate the cooperation among the different stakeholders with the reasonable use of resources to reduce the impacts of diseases (Breban et al. 2007). The disease prevention consists of three main parts, maintaining certain quantities of necessary resources to assist a prompt response, establishing the detection system of any possible outbreak and carrying out the obligatory policy to minimize potential causes which can trigger the disease. The disease control includes different response policies and the related logistics deployments.

Literature Classification and Analysis

Since different diseases have different characteristics, different prevention and control plans are needed. Therefore, we first classify the papers according to the different types of diseases. Now, most of the research papers study the smallpox, influenza (Hall 2007), HIV and anthrax. Smallpox and anthrax are the two most possible biological agents used for bioterrorist attacks (Henderson 1999). Based on the data of WHO, HIV is one of the top ten leading causes of deaths in the world recently. Influenza is a yearly outbreak epidemic (WHO Influenza (Seasonal) 2014). The high occurrence possibility of the disease is one of the main reasons why the researches of these diseases are popular. Smallpox and influenza can be transmitted by the airborne droplet, which means that people cannot realize the infection. In most of the papers, R0 and the transmission modes are the two important factors which will affect the number of calculated infected people. The value of R0 decreases after the beginning of medical intervention because the disease can be controlled by an effective medical intervention. Usually, the transmission modes are classified by homogeneity or heterogeneity. Homogeneity transmission mode means that all hosts have identical disease transmission rate while heterogeneity transmission mode means that the disease transmission rate depends on the different types of people or different infection situations. So the papers studying smallpox and influenza are classified according to how they present R0, how the transmission modes are modeled and which methods are used to solve the problem. Even though HIV is a communicable disease, HIV cannot be transmitted by the airborne droplet and so papers studying HIV are classified according to the contributions from a view of HIV prevention and control plans and the used methodologies. Because anthrax is non-communicable disease, we classify the paper studying the anthrax, using the same criteria of as HIV.

Table 1 Characteristics of papers studying smallpox

Reference	R0	Transmission mode	Methodology
Kaplan et al. (2002)	Uniform	Homogeneity	Differential equation model
Adivar and Selen (2011)	Uniform	Homogeneity	Differential/Integral equation model
Kress (2005)	Uniform	Heterogeneity	Mathematical model
Jamrog et al. (2007)	Uniform	Homogeneity	Markov chain model
Bozzette et al. (2003)	Nonuniform	Heterogeneity	Event-driven simulation
Halloran et al. (2002)	Nonuniform	Heterogeneity	Event-driven simulation
Porco et al. (2004)	Nonuniform	Heterogeneity	Event-driven simulation
Miller et al. (2006)	–	–	Event-driven simulation

Smallpox

Table 1 classifies papers studying smallpox. In Table 1, uniform means the paper supposes that the value of R0 does not change while nonuniform denotes the value of R0 changes in the study. Kaplan et al. (Kaplan et al. 2002) used the differential equation model to compare the mass vaccination and the ring vaccination after a smallpox bioterrorist attack. The service rate of a mass vaccination policy is four times higher than a traced vaccination policy. Mass vaccination policy is to vaccine all persons in the infected area. Traced vaccination policy is a strategy in which only direct contacts of confirmed victims are traced and vaccinated. Adivar and Selen (2011) proposed an integral equation model and a differential equation model to study how the quarantine policy controls smallpox. The model proposed by Kress (2005) can evaluate the effect of quarantine, public announcement and mass vaccination. Jamrog et al. (2007) studied the effect of mass vaccination, ring vaccination and combined vaccination strategy by using the Markov chain model. Bozzette et al. (2003), Halloran et al. (2002), Porco et al. (2004) and Miller et al. (2006) used the event-driven simulation to study the response to smallpox. Bozzette et al. (2003) compared the different vaccine coverage rate polices and the different individuals priority policies under various smallpox attack scenarios. Halloran et al. (2002) used a stochastic simulation model to compare mass and ring vaccination policies in a small community. Porco et al. (2004) compared the mass vaccination and ring vaccination. Porco et al. (2004) is the only paper which considers the different forms of the smallpox, which will affect the value of R0. The vaccination capacity is the most important factor to present the differences between mass vaccination and the target vaccination in Porco et al. (2004). The development of disease in Porco et al. (2004) depends on the age. Different from other papers, Miller et al. (2006) do not compare the different response policies but tries to design the response plan for San Antonio (USA). Two simulation models are proposed in this paper, the casualty prediction model and the health-care complex model (Table 2).

Table 2 Characteristics of the papers studying the influenza

Reference	R0	Transmission mode	Methodology
Arinaminpathy and McLean (2009)	Nonuniform	Heterogeneity	Mathematical model
Colizza et al. (2007)	Uniform	Heterogeneity	Mathematical model
Roberts et al. (2007)	Uniform	Heterogeneity	Mathematical model
Chen and Liao (2008)	Uniform	Homogeneity	Mathematical model
Ferguson et al. (2006)	Nonuniform	Heterogeneity	Simulation
Germann et al. (2006)	Uniform	Heterogeneity	Simulation
Milne et al. (2008)	Nonuniform	Homogeneity	Simulation

Influenza

The papers studying the influenza have been classified according to the same criteria as used by smallpox in Table 1. Arinaminpathy and McLean (2009) used the mathematical model to study how the different drug coverage rates affect the number of deaths, taking into account the limited national stockpiles. The development of disease is uniform distribution. Colizza et al. (2007) proposed a stochastic epidemic model to investigate how the travel restriction and the antivirus distribution policy affect the pandemic transmission. This model is built on a global scale, taking into account the effect of temporal and spatial characteristics on the evolution of the pandemic. The disease is assumed to follow the exponential distribution and uniform distribution in Colizza et al. (2007). Roberts et al. (2007) studies which response strategy should be adopted according to the different values of R0. Chen and Liao (2008) evaluated the effects of enhanced ventilation, use of respiratory mask and vaccination on pandemic influenza transmission in a school in Taiwan. Ferguson et al. (2006) used simulation to examine the effects of antiviral treatment, vaccination, isolation, household quarantine, school and workplace closure and travel restrictions in the United Kingdom and United States. The basic reproduction number in Ferguson et al. (2006) changes with time and the development of disease is based on the history data. Germann et al. (2006) used simulation to evaluate the different factors which may impact the number of infected patients from a strategic level. The transition rates used in Germann et al. (2006) depend on the age. Milne et al. (2008) assessed the effects of isolation, school closure and contact reduction in Australian with an individual-based stochastic simulation model. Three other interesting papers do not include so many details about the disease development. Hashikura and Kizu (2009) proposed a mathematical model to calculate the amounts of personal protective equipment required by each hospital according to the number of health care workers in the hospital and the length of pandemic period. Different from other papers, the seriousness of disease is not modeled by the basic reproduction number but the three levels: high, medium

Table 3 Characteristics of the papers studying the HIV

Reference	Contribution	Methodology
Kahn et al. (2006)	Cost-effective analysis	Linear model
Sanders et al. (2005)	Cost-effective analysis	Markov model
Zaric et al. (2000)	Cost-effective analysis	Compartmental model
Bertolli et al. (2003)	Strategies comparison	Decision tree
Bendavid et al. (1910)	Strategies comparison	Simulation
Paltiel et al. (2006)	Strategies comparison	Simulation

and low. Hessel (2009) explains systemically the challenge of the pandemic vaccine supply, the success factors for ensuring vaccine availability and the processes of the establishment of a vaccine supply chain. Harrington and Hsu (2010) tests how purchase programs used by the antiviral drugs manufactures affect the stockpiling of non-governmental organizations, such as hospitals.

HIV

Since the HIV cannot be transmitted by the airborne droplet, papers studying the HIV do not pay attention to the transmission method but the prevention policies. So, Table 3 classified the papers according to contributions from a view of HIV prevention and control plans and the used methodology. Kahn et al. (2006), Sanders et al. (2005) and Zaric et al. (2000) use the cost-effective analysis with different methods. Kahn et al. (2006) developed a linear model to calculate the costs by using male circumcision to prevent the transmission by sexual contract. This model does not have too many details about the disease transmission but this simple linear model is transparent and efficient. Sanders et al. (2005) developed a Markov model to evaluate the cost effectiveness of HIV screening strategy to control the disease transmission in the United States. The patients' behavioral has been taken into consideration. Zaric et al. (2000) used the compartmental model to estimate the cost effectiveness of using methadone maintenance treatment for injection drug users. Some authors pay more attention to the treatment results rather than the cost of different strategies. Bertolli et al. (2003) studied the strategies which can avoid the transmission from mother-to-child by a decision tree model. The decision tree model can capture the possibility of the HIV transmission from mother to child and the possibility of death if the child does not have the mother breastfeeding. Bendavid et al. (1910) and Paltiel et al. 2006 adopted the simulation to evaluate the different strategies. Bendavid et al. (1910) used the number of immune cells as an indicator to judge the different monitor strategies. Paltiel et al. (2006) evaluated how the screening strategies impact the number of the infected patients (Table 4).

Table 4 Characteristics of the papers studying the Anthrax

Reference	Contribution	Methodology
Wein et al. (2003a)	Strategies comparison	Mathematical model
Craft et al. (2005)	Response plan analysis	Mathematical model
Wein and Craft (2005)	Strategies comparison	Mathematical model
Houck and Herrmann (2011)	Strategies comparison	Compartmental model
Zaric et al. (2008)	Cost–benefit analysis	Compartmental model
Bravata et al. (2006)	Cost–effectiveness analysis	Compartmental model
Hu (2001)	Strategies comparison	System dynamics
Whitworth (2013)	Response plan evaluation	Discrete-event simulation

Anthrax

A four-paper series constitutes one of the first and most important contributions to the field. Wein et al. (2003a) proposed a mathematical model to compare five priority policies after an airborne anthrax attack. This model incorporated an atmospheric dispersion model, an age-dependent dose-infected model, a disease progression model, and an intervention policy model. Craft et al. (2005) simplified the model proposed by Wein et al. (2003a) and used the number of deaths as a key parameter to analyze the response to the anthrax bioterrorist attack. Wein and Craft (2005) used the model proposed by Wein et al. (2003a) to compare the pre-exposure prophylaxis strategy and post-exposure prophylaxis strategy. Based on the conclusions from the aforementioned papers, Wein and Kaplan (2003) identified the key elements for an effective anthrax response plan. Both Houck and Herrmann (2011) and Zaric et al. (2008) used the compartmental model. Based on the model proposed by Zaric et al. (2008), Bravata et al. (2006) evaluates the costs and lives saved by four alternative strategies. They concluded that the local dispensing capacity is the most important factor to restraint the mortality rate after anthrax bioterrorist attack. Hu and Zhao (2001) used system dynamics to analyze the different emergency response plans after an anthrax attack. Whitworth (2013) used a discrete-event simulation approach to design and evaluate the response to an anthrax attack processes. Another interesting paper in this field is by Brookmeyer et al. (2008), which focuses on the incubation period.

Based on the above analysis, two main gaps can be found. First, the number of infected individuals can be reduced by an effective vaccination policy. So the value of R0 will decrease after the medical intervention. However, most of the authors assume that the value of R0 is uniform. This assumption leads to the number of the calculated infected individuals which is a little far from the real world situation. Second, the outbreak of disease is always a large scale disaster. However, most of studies are done based on a small scale scenario.

Discussions and Future Directions

Discussions

From the classification and above analysis, it can be found that mathematical model is one of the most popular methods. In order to present the propagation of diseases, the compartmental model has been used as a foundation for a lot of mathematical model in this field. The compartmental model can reduce the population diversity to a few key characteristics which can be used to study the progress of an epidemic and the intervention policies easily. The compartmental model stratifies the populations into several stages or compartments. All of these compartments are mutually exclusive. The compartmental model is easy to refine or extend when different medical resources are taken into account. To project the development of different stages over time, most of the compartmental models are dynamic models. At the beginning of twentieth century, Kermack and McKendrick (1932) proposed the Susceptible-Infectious-Recovered (SIR) model which is based on the principle of compartmental model. The SIR model divides the people according to the health status. The SIR model is always a nonlinear model and is used to study the communicable diseases. In other words, SIR model is a special kind of compartmental model which can present the disease development intuitively. For both compartmental model and SIR model, the disease development and the medical intervention can be connected by some key important factors of medical intervention, such as the coverage rate of vaccines. This connection can help the decision makers to analyze how these factors affect disease progression easily. However, the mathematical model is difficult in building a complete model which can reflect reality. For example, some mathematical models suppose the homogeneity transition instead of the heterogeneity transmission mode which is not true in the real world. Moreover, an accurate model, which includes too many details, needs an effective algorithm and requires enormous computer time. The computation difficulty is another limitation.

Compared with mathematical model, simulation can be much easier to reflect the situation in the real world. From a view of disease development, simulation can capture the detail of the real situation, such as the change of R0 and the heterogeneity transmission mode. From a view of disease prevention, simulation is suitable to handle a large scale problem, such as the disease prevention plans for more than one million individuals. However, the connection between the development of the disease and the medical intervention is not explicit. This inherent connection will make the barrier for the numerical analysis about how the disease prevention affects the disease development.

Future Research Directions

Following the insights provided above, there are a lot of opportunities for future research efforts. From a view of disease, first, lots of models failed to combine some factors which can impact the number of casualties, such as the different periods in which the patients transfer into the current disease stage. Second, not all the papers studying the response to the diseases were based on real cases and the real emergency management plans. For example, most of the authors assume the recovery rate remains the same for the patients in the same stage while in reality, the recovery rate will decrease when the patients stay in the stage longer. Third, the number of infected individuals will affect the adopted response policy. If the development of the disease follows different disease development probability functions, the number of infected individuals will change. But many authors failed to consider about how disease development probability functions affect the medical interventions.

From a view of disease prevention and control plans, first, the side effect of the medical interventions are neglected by most of the authors. For example, the side effect of the mass vaccination of smallpox has often been neglected. Second, some key important logistics factors, such as the distribution capacity, are not taken into account. The medical intervention can work well only when the medical resources can be delivered in time with the right amount. Though some papers take, into account, the limitation of the medical treatment capacity, they do not address logistics questions like the number and the size of the antibiotic distribution centres. Moreover, most of the papers consider only one factor, which may affect the medical intervention, such as the vaccination coverage rate, but neglect the national stockpile. However, the sudden occurrence of epidemic may be more transmissible than we predicted and more people will be infected. So the national stockpile may be exhausted and a lot of individuals cannot get the medical help in time. In other words, the vaccination coverage policy cannot be executed well without enough medical resources. So, how to dispense medical resources sparingly to avoid exhausting the stockpile should be studied as well. Third, some individuals, such as the old and the young, have the high possibilities to be infected and need special help. But most of the authors focus their attention on the general population, and neglected them who need the special help. Fourth, it can be found that most of the papers study the vaccination policy because vaccination is one of the most effective ways to prevent and control disease. But the quarantine policy, isolation policy and so on should also get enough attention.

References

Adivar, B., Selen, E.S.: Compartment disease transmission models for smallpox. Discrete Cont. Dyn. Syst. 13–21(2011)

Arinaminpathy, N., McLean, R.: Logistics of control for an influenza pandemic. Epidemics **1**(2), 83–88 (2009)

Bendavid, E., Young, S.D., Katzenstein, D.A., Bayoumi, A.M., Sanders, G.D., Owens, D.K.: Cost-effectiveness of HIV monitoring strategies in resource-limited settings: a southern African analysis. Arch. Intern. Med. **168**(17), 1910–1918 (2008).

Bertolli, J., Hu, D.J., Nieburg, P., Macalalad, A., Simonds, R.J.: Decision analysis to guide choice of interventions to reduce mother-to-child transmission of HIV. AIDS **17**(14), 2089–2098 (2003)

Bozzette, S., Boer, R., Bhatnagar, V., Brower, J.L., Keeler, E.B., Morton, S.C., Stoto, M.: A model for a smallpox-vaccination policy. N. Engl. J. Med. **348**, 416–425 (2003)

Bravata, D.M., Zaric, G.S., Holty, J.C., Brandeau, M.L., Wilhelm, E.R., McDonald, K.M., Owens, D.K.: Reducing mortality from anthrax bioterrorism: strategies for stockpiling and dispensing medical and pharmaceutical supplies. Biosecur. Bioterror. **4**, 244–262 (2006)

Breban, R., Vardavas, R., Blower, S.: Theory versus data: how to calculate R0? PLoS One **2**(3), e282 (2007)

Brookmeyer, R., Johnson, E., Barry, S.: Modeling the incubation period of anthrax. Stat. Med. **24**, 531–542 (2008)

Chen, S.C., Liao, C.M.: Modelling control measures to reduce the impact of pandemic influenza among schoolchildren. Epidemiol. Infect. **136**, 1035–1045 (2008)

Coburn, B.J., Wagner, B.G., Blower, S.: Modeling influenza epidemics and pandemics: insights into the future of swine flu (H1N1). BMC Med. **7**(1), 30 (2009)

Colizza, V., Barrat, A., Barthelemy, M., Valleron, A.-J., Vespignani, A.: Modeling the worldwide spread of pandemic influenza: baseline case and containment interventions. PLoS Med. **4**(1), e13 (2007)

Craft, D.L., Wein, L.M., Wilkins, A.H.: Analyzing bioterror response logistics: the case of anthrax. Manag. Sci. **51**, 679–694 (2005)

Dasaklis, T.K., Pappis, C.P., Rachaniotis, N.P.: Epidemics control and logistics operations: a review. Int. J. Prod. Econ. **139**(2), 393–410 (2012)

Ferguson, N.M., Cummings, D.T., Fraser, C., Cajka, J.C., Cooley, P.C., Burke, D.S.: Strategies for mitigating an influenza pandemic. Nature **442**(7101), 448–52 (2006)

Germann, T.C., Kadau, K., Longini, I.M., Macken, C.: Mitigation strategies for pandemic influenza in the United States. Proc. Natl. Acad. Sci. U. S. A. **103**(15), 5935–5940 (2006)

Hall, C.B.: The spread of influenza and other respiratory viruses: complexities and conjectures. Clin. Infect. Dis. **45**(3), 353–359 (2007)

Halloran, M.E., Longini, I.M., Nizam, A., Yang, Y.: Containing bioterrorist smallpox. Science **298**(5597), 1428–1432 (2002)

Harrington, J.E., Hsu, E.B.: Stockpiling anti-viral drugs for a pandemic: the role of Manufacturer Reserve Programs. J. Health Econ. **29**(3), 438–544 (2010)

Hashikura, M., Kizu, J.: Stockpile of personal protective equipment in hospital settings: preparedness for influenza pandemics. Am. J. Infect. Control **37**, 703–707 (2009)

Henderson, D.A.: Biological terrorism—The looming threat of bioterrorism. Science **283**, 1279–1282 (1999)

Hessel, L.: Pandemic influenza vaccines: meeting the supply, distribution and deployment challenges. Influenza Other Respir. Viruses **3**(4), 165–70 (2009)

Houck, M.L., Herrmann, J.W.: Preparing for an anthrax attack: the impact of distributing MedKits. IERC, Arkansas (2011)

Hu, J., Zhao, L.: Emergency logistics strategy in response to anthrax attacks based on system dynamics. Int. J. Math. Operat. Res. **3**, 490–509 (2001)

Jamrog, D.C., Shatz, M.P., Smith, C.: Modeling responses to anthrax and smallpox attacks. Lincoln Lab. J. **17**(1), 115–129 (2007)

Kahn, J.G., Marseille, E., Auvert, B.: Cost-effectiveness of male circumcision for HIV prevention in a South African setting. PLoS Med. **3**(12), e517 (2006)

Kaplan, E.H., Craft, D.L., Wein, L.M.: Emergency response to a smallpox attack: the case for mass vaccination. Proc. Natl. Acad. Sci. U. S. A. **99**(16), 10935–40 (2002)

Kermack, W.O., McKendrick, A.G.: Contributions to the mathematical theory of epidemics—II. The problem of endemicity. 1932. Bull. Math. Biol. **138**(834), 55–83 (1932)

Kress, M.: The effect of social mixing controls on the spread of smallpox—a two-level model. Health Care Manag. Sci. **8**(4), 277–289 (2005)

Long, E.F., Vaidya, N.K., Brandeau, M.L.: Controlling co-epidemics: analysis of HIV and tuberculosis infection dynamics. Oper. Res. **56**(6), 1366–1381 (2008)

Miller, G., Randolph, S., Patterson, J.E.: Responding to bioterrorist smallpox in San Antonio. Interfaces **36**(6), 580–590 (2006)

Milne, G.J., Kelso, J.K., Kelly, H.A., Huband, S.T., McVernon, J.: A small community model for the transmission of infectious diseases: comparison of school closure as an intervention in individual based models of an influenza pandemic. PLoS One **3**, e4005 (2008)

Paltiel, D., Weinstein, M.C., Kimmel, A.D., Seage, G.R., Losina, E., Zhang, H., Walensky, R.P.: Expanded screening for HIV in the United States-an analysis of cost-effectiveness. N. Engl. J. Med. **352**, 586–595 (2006)

Porco, T.C., Holbrook, K.A., Fernyak, S.E., Portnoy, D.L., Reiter, R., Aragón, T.J.: Logistics of community smallpox control through contact tracing and ring vaccination: a stochastic network model. BMC Public Health **4**, 1–34 (2004)

Roberts, M.G., Baker, M., Jennings, L.C., Sertsou, G., Wilson, N.: A model for the spread and control of pandemic influenza in an isolated geographical region. J. R. Soc. Interface **4**(13), 325–30 (2007)

Sanders, G.D., Bayoumi, A.M., Sundaram, V., Bilir, S.P., Neukermans, C.P., Rydzak, C.E., Douglass, L.R.: Cost-effectiveness of screening for HIV in the era of highly active antiretroviral therapy. N. Engl. J. Med. **352**(6), 570–585 (2005)

Wein, L.M., Craft, D.L.: Evaluation of public health interventions for anthrax: a report to the secretary's council on public health preparedness. Biosecur. Bioterror. **3**, 348–356 (2005)

Wein, L.M., Craft, D.L., Kaplan, E.H.: Emergency response to an anthrax attack. Proc. Natl. Acad. Sci. U. S. A. **100**, 4346–4351 (2003a)

Wein, L.M., Kaplan, E.H.: Unready for anthrax. The Washington Post, July 28 (2003).

Whitworth, M.H.: Designing the response to an anthrax attack. Interfaces **36**, 562–568 (2013)

WHO Global Influenza Preparedness Plan. The role of WHO and recommendations for national measures before and during pandemics. World Health Organization (2005)

WHO Influenza (Seasonal) Fact sheet N°211. March 2014.

WHO—Affected Countries Ebola Virus Disease Outbreak Response Plan in West Africa. World Health Organization, 31 July 2014

Zaric, G.S., Barnett, P.G., Brandeau, M.L.: Methadone maintenance and HIV prevention: a cost-effectiveness analysis. Manag. Sci. **46**(8), 1013–1031 (2000)

Zaric, G.S., Bravata, D.M., Cleophas, H.J.E., McDonald, K.M., Owens, D.K., Brandeau, M.L.: Modeling the logistics of response to anthrax bioterrorism. Med. Decis. Making **28**, 332–350 (2008)

A Goal-Programming Approach to the Master Surgical Scheduling Problem

Paola Cappanera, Filippo Visintin, and Carlo Banditori

Introduction

The surgical scheduling is the process that aims to plan the activities of the operating theatre. Its economic and social impact together with its complexity have stimulated in the recent years intensive research activities (Cardoen et al. 2010; Guerriero and Guido 2011; May et al. 2011). In the literature, such a process is usually seen as a three-stage cascade process, in which the output of the upstream sub-process represents the input for the downstream one (Beliën and Demeulemeester 2007): (1) the case mix planning, (2) the master surgical scheduling (MSS) and (3) the patient selection and sequencing. This study focuses on the second stage, i.e. the MSS, which consists in the determination of the surgical specialty and/or the typology of surgeries to assign to each OR, in each session, in each day, i.e. to each OR session, of a given planning horizon. In general, solving such a problem is noticeably complex. In fact, it requires to consider: (1) many different types of cases, characterised by different priority levels and requiring different surgical procedures; (2) many different types of resources, such as ORs, OR personnel (e.g., surgeons, anaesthetists and nurses), surgical and electro-medical equipment, post-surgical resources (e.g., ICU, post-surgical units); (3) the randomness associated with patients' arrival, surgeries' duration and patients' length of stay (LoS) (May et al. 2000); and (4) the conflicting priorities and preferences of the scheduling process stakeholders (Glouberman and Mintzberg 2001). These stakeholders are:

P. Cappanera (✉)
IBIS Lab, Dipartimento di Ingegneria dell'Informazione, Università degli Studi di Firenze, Firenze, Italy
e-mail: paola.cappanera@unifi.it

F. Visintin • C. Banditori
IBIS Lab, Dipartimento di Ingegneria Industriale, Università degli Studi di Firenze, Firenze, Italy

A. Matta et al. (eds.), *Health Care Systems Engineering for Scientists and Practitioners*, Springer Proceedings in Mathematics & Statistics 169,
DOI 10.1007/978-3-319-35132-2_15

(1) the patient, who wants his/her priority to be respected and thus his/her surgery to be scheduled before the assigned latest due date (LDD); (2) the OR staff (surgeons, anaesthetists, nurses), who wants a fair distribution of the OR workload; (3) the hospital management, whose aim is to increase the OR productivity in order to contain the associated costs. In order to fulfil the stakeholders' expectations, an ideal MSS should be efficient, balanced, robust (Banditori et al. 2014) and should be created considering the priorities and LDDs of the patients in the waiting list. However, these objectives can be conflicting and consequently the decision maker is supposed to find a solution representing a good trade-off.

In this study, we present a novel mixed-integer programming model which complies with the stakeholders' requirements considering multiple objectives, i.e. the maximisation of the number of scheduled surgeries, the patients' LDDs fulfilment and the ORs and-post surgical units utilisation. This four-criteria problem is addressed by means of a goal programming approach in which the relative importance of each criterion can be fixed by tuning a set of weights. In this paper, we present the preliminary results of numerical experiments aiming at testing the model and at showing its ability to provide solutions that can fit different requirements. The numerical experiments are based on real data coming from a leading Italian hospital.

Literature Review

The literature abounds of contributions in the OR planning and scheduling field (Cardoen et al. 2010; Guerriero and Guido 2011). Great attention has been dedicated also to the MSS problem. A broad re-view about the MSS problem is out of the scope of this paper. The interested reader is referred to (Banditori et al. 2013; Cappanera et al. 2014; Visintin et al. 2014). In this section, we focus on the papers proposing multi-criteria models supporting the MSS process.

In (van Oostrum et al. 2008) the authors propose a bi-criteria optimisation model in which the objective function minimises the OR capacity and levels the beds occupation over the planning horizon. Beds levelling is also considered in Beliën et al. (2009) where, in addition, the objective function aims to reduce the number of ORs shared by different surgical specialties and to make the schedule as much repetitive as possible. In (Adan et al. 2009), instead, the objective function minimises the deviations of the resources utilisation from fixed targets. Specifically the model takes into consideration the ORs, two types of beds, i.e. medium and intensive care, and intensive care ward nurses.

Finally, in (Rachuba and Werners 2013) the authors propose a MIP model in which the objective function considers the patients waiting time, the OR overtime and the number of patients that must be deferred to the next planning horizon because of lack of capacity. Hence, despite efficiency, resource balancing and patients' priorities are by no means novel optimisation criteria used in the production of the MSS, to the best of our knowledge, no model integrates all of them in a single objective function. With this paper, we specifically address this literature gap.

Problem Addressed

In this section, we describe the main features characterising the MSS problem addressed in this study. We assume that scheduling surgeries requires the simultaneous availability of three resources, namely (1) the ORs, which are totally interchangeable and whose available time is subdivided in sessions, e.g. daily, morning or afternoon; (2) the post-surgical beds (hereinafter beds) that are organised in different units; (3) the surgeons teams, each of which belongs to a single surgical specialty. With regard to this latter point we assume that surgical specialties are assigned to OR sessions a priori by means of an allocation grid. Such a grid indicates which surgical specialty will operate in each OR and session on each day of the planning horizon, according to the availability of the surgeons teams. We do not consider other resources, such as electro-medical equipment or nursing staff, since we hypothesise them not scarce and thus not critical for the scheduling purposes.

We assume that the patients in the hospital waiting lists can be categorised in the so-called *surgery groups*. Patients belonging to the same surgery group require a surgeon team of the same specialty, are expected to require a similar amount of OR time, i.e. are characterised by a similar expected surgical time (ST), and are expected to occupy a bed for the same number of days, i.e. are characterised by the same expected length of stay (LoS). Apart from the surgery groups, we assume that cases in the waiting lists are characterised by the following attributes: (1) priority class, which indicates the number of days within which the patient should be scheduled, thereby defining the relevant LDD, (2) the earliest programmable date (EPD), which indicates the first day from which the patient is eligible to be scheduled. EPD is assumed equal to the first day of the planning horizon for all those cases whose EPD occurred before the current planning horizon.

Considering these issues, the problem is to determine the number of cases belonging to each surgery group to schedule in each OR session in the planning horizon, with the aims to:

- schedule as many patients as possible before the relevant LDDs;
- balance the daily utilisations of the ORs and the bed units;
- maximise the number of scheduled surgeries.

Model Description

In order to address the specific MSS problem presented in the previous section, we formulated a mixed integer goal-programming model. The model comprises the following sets and parameters:

S	Set of surgical specialties
K	Set of surgery groups
O	Set of ORs
D	Set of days in the planning horizon
T	Set of sessions
B	Set of beds units
P	Set of patients' priority classes
I	Set of criteria
G_{sodt}	The allocation grid, equal to 1 if surgeries belonging to the specialty s can be performed in OR o, on day d, in session t, 0 otherwise
s_k	Specialty of surgery group k
H_{odt}	Available time of OR o on day d, time slot t
γ_k	Expected ST of surgery group k
α_k	Expected LoS after surgery required by group k
β_k	Expected LoS before surgery required by group k
R_{bd}	Number of beds in unit b available on day d
E_{kd}	Number of cases in surgery group k, whose EPD is on day d
e	Number of time periods preceding and following the planning horizon
L_{pd}	Number of cases of priority p whose LDD is on day $d, \forall p \in P, \forall d \in D \cup D_1 \cup \cdots \cup D_e$
$\widehat{q}$	Target value for the OR utilisation rate
$\widehat{r}$	Target value for the bed utilisation rate
w_{pj}	Penalty associated with cases of priority p with LDD in D_j not scheduled in the planning horizon, $\forall p \in P, \forall j \in 1..e$
$\widehat{n_i}$	Target value for the objective i
W_i	Weight associated with criteria i

Let us introduce the decision variables:

y_{kpodt}	Number of procedures of surgery group k with priority p assigned to OR o on day d in time slot t

z_{bd}	Number of beds of type b occupied on day d
q_{odt}	Utilisation rate of OR o, on day d, in session t
r_{bd}	Utilisation rate of beds in unit b on day d
u_{pj}	Number of cases with priority p and with LDD in time period D_j not scheduled in the planning horizon, $\forall p \in P, \forall j \in 1..e$

and the auxiliary variables that are described in Table.

q^+_{odt}	Positive deviation of the utilisation rate of OR o, on day d, in session t from the fixed target
q^-_{odt}	Negative deviation of the utilisation rate of OR o, on day d, in session t from the fixed target
r^+_{bd}	Positive deviation of the utilisation rate of beds in unit b on day d from the fixed target
r^-_{bd}	Negative deviation of the utilisation rate of beds in unit b on day d from the fixed target
n_i	Value associated with objective i
n^+_i	Positive deviation of objective i from the fixed target
n^-_i	Negative deviation of objective i from the fixed target

Since the allocation grid is fixed in input through the parameter G_{sodt}, the surgery groups that can be scheduled in a given OR o, on a given day d in the time slot t are restricted to the ones belonging to the surgical specialty s for which the parameter G_{sodt} is equal to 1. For this reason variables y are defined on a subset of the set $(K \times P \times O \times D \times T)$. Specifically, we introduce, for each surgery group k, the set A_k that is a collection of triples (o,d,t) indicating, the OR sessions, in which the surgery group k can be scheduled. More formally, if s_k denote the specialty of surgery group k, A_k are defined as follows:

$$A_k = \{(o,d,t)\; s.t. o \in O, d \in D, t \in T \quad \text{and} \quad G_{s_k odt} = 1\}$$

Variables y_{kpodt} are thus defined $\forall k \in K, \forall p \in P, \forall\, (o,d,t) \in A_k$. Given these sets, parameters and variables, the model is formulated as follows:

$$\sum_{\substack{k \in K, p \in P: \\ (o,d,t) \in A_k}} \gamma_k y_{kpodt} \le H_{odt} \quad \forall o \in O, \forall d \in D, \forall t \in T \tag{1}$$

$$\sum_{\substack{k \in K, p \in P, o \in O, \\ t \in T: (o,d,t) \in A_k}} \sum_{d'=\max(1,d-\alpha_k)}^{\min(|D|,d+\beta_k)} y_{kpod't} = z_{bd} \quad \forall b \in B, \forall d \in D \tag{2}$$

$$z_{bd} \le R_{bd} \quad \forall b \in B, \forall d \in D \tag{3}$$

$$q_{odt} = \frac{\sum_{\substack{k \in K, p \in P: \\ (o,d,t) \in A_k}} \gamma_k y_{kpodt}}{H_{odt}} \quad \forall O \in O, \forall d \in D, \forall t \in T \tag{4}$$

$$r_{bd} = \frac{z_{bd}}{R_{bd}} \quad \forall b \in B, \forall d \in D \tag{5}$$

$$\sum_{\substack{p \in P, o \in O, d' \in D: d' \le d, \\ t \in T: (o,d',t) \in A_k}} y_{kpod't} \le \sum_{d' \in D: d' \le d} E_{kd'} \quad \forall k \in K, \forall d \in D \tag{6}$$

$$\sum_{\substack{k \in K, o \in O, d \in D, \\ t \in T : (o,d,t) \in A_k}} y_{kpodt} + \sum_{h=1}^{j} u_{ph} \geq \sum_{d \in D \cup D_1 \cup \cdots \cup D_j} L_{pd} \quad \forall p \in P, \forall j \in 1..e \tag{7}$$

$$q^+_{odt} \geq q_{odt} - \widehat{q} \quad \forall o \in O, \forall d \in D, \forall t \in T \tag{8}$$

$$q^-_{odt} \geq \widehat{q} - q_{odt} \quad \forall o \in O, \forall d \in D, \forall t \in T \tag{9}$$

$$r^+_{bd} \geq r_{bd} - \widehat{r} \quad \forall b \in B, \forall d \in D \tag{10}$$

$$r^-_{bd} \geq \widehat{r} - r_{bd} \quad \forall b \in B, \forall d \in D \tag{11}$$

$$n_1 = \sum_{p \in P, j \in 1..e} w_{pj} u_{pj} \tag{12}$$

$$n_2 = \sum_{\substack{o \in O, d \in D, \\ t \in T}} q^+_{odt} + q^-_{odt} \tag{13}$$

$$n_3 = \sum_{b \in B, d \in D} r^+_{bd} + r^-_{bd} \tag{14}$$

$$n_4 = \sum_{\substack{k \in K, p \in P, o \in O, \\ d \in D, t \in T : (o,d,t) \in A_k}} y_{kpodt} \tag{15}$$

$$n^+_i \geq n_i - \widehat{n}_i \quad \forall i \in I \tag{16}$$

$$n^-_i \geq \widehat{n}_i - n_i \quad \forall i \in I \tag{17}$$

$$y_{kpodt} \in \mathrm{N} \quad \forall k \in K, \forall p \in P, \forall (o,d,t) \in A_k \tag{18}$$

$$u_{pj} \geq 0 \quad \forall p \in p, \forall j \in 1.e \tag{19}$$

$$z_{bd} \geq 0, \ r^+_{bd} \geq 0, \ r^-_{bd} \geq 0 \quad \forall b \in B, \forall d \in D \tag{20}$$

$$q^+_{odt} \geq 0, \ q^-_{odt} \geq 0 \quad \forall o \in O, \forall d \in D, \forall t \in T \tag{21}$$

$$n^+_i \geq 0, \ n^-_i \geq 0 \quad \forall i \in I \tag{22}$$

$$\min \sum_{i \in I} \frac{W_i \left(n_i^+ + n_i^-\right)}{\widehat{n_i}} \tag{23}$$

Constraints (1) assure that for each OR session, the sum of the STs of the scheduled surgeries does not exceed the available time. Constraints (2) compute the number of utilised beds for each unit and for each day of the planning horizon. Constraints (3) limit the number of occupied beds. Constraints (4) and constraints (5) compute respectively the daily utilisation of the OR sessions and of the different units. Constraints (6) assure that the number of scheduled surgeries for each group does not exceed the number of cases that are available, depending on the relevant EPDs.

Constraints (7) allow for the respect of the LDDs' of the patients in the waiting list. Specifically, these covering constraints impose that the number of scheduled surgeries of a given priority p should be greater or equal than the number of cases in the waiting lists belonging to that priority. If this cannot happen the corresponding variable u, which measures the number of not scheduled surgeries of priority p and with a LDD falling in the time period D or D_j, assumes a value greater than zero. The sum of the u variables is penalised in the objective n_1 according to the priority and the tardiness.

Constraints (8) and (9) compute positive and negative deviations of OR utilisation from the fixed target, for each triple (o,d,t). Constraints (10) and (11) are their counterparts for the unit utilisation. Constraints (12)–(15) compute the values of the four criteria: specifically, constraint (12) computes the weighted sum of the penalties associated with the missed scheduling of the patients with certain LDDs and priorities; constraints (13) and (14) calculate the sum of the deviations from the fixed targets respectively of ORs and units; constraint (15) computes the number of scheduled surgeries. Constraints (16) and (17) compute, for each criterion the positive and the negative deviation from the fixed target. The remaining constraints define variable domains. The weighted sum of the deviations calculated in constraints (16) and (17) is minimised in the objective function (23).

Numerical Experiments

In this section, we describe the preliminary numerical experiments we carry out in this study and present the relevant results. Firstly, we introduce the data the experiments are based on. Secondly, we illustrate how the experiments were designed. Finally, we show the results of the study. The model has been coded in AMPL and solved through the solver IBM ILOG CPLEX (version 12.4).

Input Data

The numerical experiments are based on real data coming from one leading Italian hospital. The model performance has been tested in correspondence with ten different instances. Each instance differs from the others in terms of composition of the waiting list, i.e. each instance is characterised by a different set of patients to be scheduled and thus by different surgery groups, EPDs and LDDs combinations. Specifically we have considered:

- a planning horizon of 2 weeks;
- 11 surgical specialties and a number of surgery groups ranging from 67 to 75 (depending on the particular instance);
- 4 ORs;
- 3 beds units, i.e. day surgery unit, which accommodates the patients with LoS equal to one day independently from specialty, week hospital and ordinary unit, accommodating the patients with LoS higher than one day of specific specialties;
- a target value for the daily utilisations of OR sessions and bed units of 85 %;
- in order to penalise the missing scheduling within their LDDs, patients are clustered in four time periods ($e = 3$), according to the relevant LDDs:
  - patients whose LDD is in the planning horizon D;
  - D_1, patients whose LDD is expired from a number of day smaller than the number of days of the planning horizon;
  - D_2, patients whose LDD is expired from a number of day bigger than the number of days of the planning horizon;
  - D_3, patients whose LDD occurs after the planning horizon.

The weights w_{pj} in the objective n_1 are set in the way that the model prioritises (1) cases belonging to different LDDs' time clusters; and (2) cases with higher priority. With respect to the time clusters the weights are set in the way that the model priorities the patients whose LDD expires in the planning horizon (D), then it chooses among the patients whose LDD is expired from less than 14 days (D_1) and then more than 14 days (D_2); finally it chooses among the patients whose LDDs expires beyond the planning horizon (D_3);

- the targets for the different objectives are fixed as the ideal values that these objectives can assume; specifically they are calculated by solving the models considering only one objective at a time. The values obtained solving the four sub-problems are then used to solve the goal programming model. Since the target values are ideal values, depending on the typology of objective that is considered, i.e. minimisation or maximisation, respectively negative and positive deviation variables with respect to the objective can be eliminated.

Experimental Design

In order to test the ability of the model to produce satisfactory solutions, it is necessary to analyse the impact of different combinations of the weights associated with the four objectives. Jones and Tamiz (2010) propose an algorithm to generate a set of weight combinations to systematically explore the weight space. The algorithm is based on two parameters: *Tmax*, which represents the number of weights that are varied simultaneously, and *max_level*, that indicates the level of granularity adopted in the exploration of the weight space. The algorithm is initialised with an initial set of weights. In our numerical experiments we adopted *Tmax* = 2 and *max_level* = 1. Starting from the initial combination (25 %, 25 %, 25 %, 25 %) we obtained the 21 different weight combinations. It is worth to observe that deviations of a solution from the target values, per se, do not allow understanding if the solution should be considered acceptable. For example, if we set a target of two for the beds utilisation criterion, a 100 % deviation from the target might be considered perfectly acceptable by the hospital management (indeed, it implies having two vacant beds for one day in the whole planning horizon). As a consequence, to assess the acceptability of a solution we set a threshold value for each criterion. Hence, with respect to the identified criterion, we consider "acceptable" those solutions that return values smaller than the threshold. The threshold values were set based on recommendations from the hospital management.

Experimental Results

Table 1 reports the results of the numerical experiences. For each of the 21 weight combinations we calculated the mean deviations across the ten instances of the objective values from the fixed targets and the solution acceptance rate, i.e. the number of times that we obtained values under the acceptance threshold.

As an example $n^{+}_{LDD} = 51.1$ in the first row of the table (Combs = 1) means that the value of the objective function that measures the penalty resulting from the latest due date violation (first criterion) exceeds, on average, by 51.1 % the corresponding target value, where the mean is calculated over the ten instances.

Looking at Table 1 it can be noticed that none of the combinations produces solutions that are acceptable with respect to all the criteria. However, some combinations lead to a good trade-off among the four criteria. For example, let us consider combinations 11 and 16. These combinations led to acceptable solutions in terms of OR balancing, bed balancing and scheduled surgeries, for all the instances. In addition, they led to satisfactory solutions also for the LDDs criterion in six and five out of ten instances, respectively.

Table 1 Experimental results

	Weights [%]				Mean deviations [%]				Solution acceptance rate			
Combs'	W_{LDD}	W_{OR}	W_{Bed}	W_N	n^+_{LDD}	n^+_{OR}	n^+_{Bed}	n^-_N	LDD	OR	Bed	N
1	25	25	25	25	51.1	91	105	20.8	4	10	10	10
2	12.7	62.0	12.7	12.7	70.1	44.6	190	21.4	1	10	9	7
3	0.3	99	0.3	0.3	71.7	14.7	1275	18.6	2	10	0	10
4	12.7	12.7	62	12.7	49.9	313.4	10	19.4	4	1	10	10
5	0.3	0.3	99	0.3	52	314.3	10	19.6	7	1	10	10
6	62	12.7	12.7	12.7	35.6	95.1	140	21.4	6	10	10	8
7	99	0.3	0.3	0.3	9.5	758.8	765	17.9	9	1	2	9
8	12.7	12.7	12.7	62	52.8	77.2	125	20.5	5	10	10	10
9	0.3	0.3	0.3	99	83.8	307.3	2020	2.4	0	1	0	10
10	49.5	0.5	0.5	49.5	9.5	469	850	14.2	10	0	0	10
11	37.3	12.8	12.8	37.3	39	92.2	130	21.1	6	10	10	10
12	12.8	37.3	37.3	12.8	64.9	105.6	95	21.2	1	9	10	9
13	0.5	49.5	49.5	0.5	79.5	101.1	95	21.6	0	9	10	6
14	0.5	0.5	49.5	49.5	61.4	452.6	0	14.8	3	0	10	10
15	12.8	12.8	37.3	37.3	47.6	315.3	10	18.7	5	2	10	10
16	37.3	37.3	12.8	12.8	43.4	62.9	160	21	5	10	10	10
17	49.5	49.5	0.5	0.5	28.8	106.8	1340	17.1	9	9	0	10
18	49.5	0.5	49.5	0.5	13.9	904.3	0	20.3	10	0	10	8
19	37.3	12.8	37.3	12.8	38.5	314.8	15	19.6	6	3	10	10
20	12.8	37.3	12.8	37.3	62.1	58.1	155	21	3	10	10	9
21	0.5	49.5	0.5	49.5	70.2	56.1	1290	14	2	9	0	10

Looking at the values relevant to the single criteria, we can observe:

- with respect to the LDDs criterion, extremely high values of acceptance rate, e.g. 90–100 %, are obtained when the associated weight is higher than 49.5 %; however acceptance rate higher than the 50 % are obtained also with very low weights;
- very high acceptance rates are achieved with lower weights in the other cases; this is particularly true for the number of scheduled surgeries, in which a 100 % of acceptance rate is achieved also when the weight is set at the 0.3 %.

Conclusions and Future Research

In this paper, we present a novel mixed integer goal-programming model supporting the master surgical scheduling. The model takes into account four criteria that according to the literature and to the know-how we acquired in previous studies (Banditori et al. 2013, 2014; Cappanera et al. 2014; Visintin et al. 2014) capture the main peculiarities of the problem in several real-life settings. They are the followings: (1) the respect of the patients LDDs, (2) the balancing of the OR utilisation, (3) the balancing of the post-surgical units utilisation and (4) the maximisation of the number of scheduled surgeries. The importance of these criteria can be set through a set of weights. The presented numerical experiments are based on real data and comprise the study of 21 different weight combinations and ten instances. The results show that the model is able to produce solutions suitable to match different stakeholders' preferences. However, the results suggest that the impact of one weight on the different performances is not easy to determine a priori and strongly depends on the context under study. Differently from (Banditori et al. 2014), in this study the stochasticity of surgery times and lengths of stays is not addressed; the determination of robust solutions thus identifies a future line of research. Further research is also needed to explore additional settings and improve the external validity of our findings.

References

Adan, I., Bekkers, J., Dellaert, N., Vissers, J., Yu, X.: Patient mix optimisation and stochastic resource requirements: a case study in cardiothoracic surgery planning. Health Care Manag. Sci. **12**(2), 129–141 (2009)

Banditori, C., Cappanera, P., Visintin, F.: A combined optimization–simulation approach to the master surgical scheduling problem. IMA J. Manag. Math. **24**(2), 155–187 (2013)

Banditori, C., Cappanera, P., Visintin, F.: Investigating the relationship between resources balancing and robustness in master surgical scheduling. In: Matta, A., Li, J., Sahin, E., Lanzarone, E., Fowler, J. (eds.) Proceedings of the International Conference on Health Care Systems Engineering. Springer, Berlin (2014)

Beliën, J., Demeulemeester, E.: Building cyclic master surgery schedules with leveled resulting bed occupancy. Eur. J. Oper. Res. **176**(2), 1185–1204 (2007)

Beliën, J., Demeulemeester, E., Cardoen, B.: A decision support system for cyclic master surgery scheduling with multiple objectives. J. Sched. **12**(2), 147–161 (2009)

Cappanera, P., Visintin, F., Banditori, C.: Comparing resource balancing criteria in master surgical scheduling: A combined optimisation-simulation approach. Int. J. Prod. Econ. **158**, 179–196 (2014)

Cardoen, B., Demeulemeester, E., Beliën, J.: Operating room planning and scheduling: A literature review. Eur. J. Oper. Res. **201**(3), 921–932 (2010)

Glouberman, S., Mintzberg, H.: Managing the care of health and the cure of disease-Part I: differentiation. Health Care Manag. Rev. **26**(1), 56–69 (2001)

Guerriero, F., Guido, R.: Operational research in the management of the operating theatre: a survey. Health Care Manag. Sci. **14**(1), 89–114 (2011)

Jones, D., Tamiz, M.: Practical Goal Programming, vol. 141. Springer, Heidelberg (2010)

May, J.H., Strum, D.P., Vargas, L.G.: Fitting the lognormal distribution to surgical procedure times. Decis. Sci. **31**(1), 129–148 (2000)

May, J.H., Spangler, W.E., Strum, D.P., Vargas, L.G.: The surgical scheduling problem: current research and future opportunities. Prod. Oper. Manag. **20**(3), 392–405 (2011)

Rachuba, S., Werners, B.: A robust approach for scheduling in hospitals using multiple objectives. J. Oper. Res. Soc. **65**(4), 546–556 (2013) van Oostrum, J.M., Van Houdenhoven, M., Hurink, J., Hans, E., Wullink, G., Kazemier, G.: A master surgical scheduling approach for cyclic scheduling in operating room departments. OR Spectr. **30**(2), 355–374 (2008)

Visintin, F., Cappanera, P., Banditori, C.: Evaluating the impact of flexible practices on the master surgical scheduling process: an empirical analysis. Flex. Serv. Manuf. J. **1–24** (2014)

How Do Missing Patients Aggravate Emergency Department Overcrowding? A Real Case and a Simulation Study

Yong-Hong Kuo, Janny M.Y. Leung, and Colin A. Graham

Introduction

Emergency department (ED) overcrowding refers to the situation that emergency medicine service is obstructed because the number of patients waiting to be seen, undergoing assessment and medical treatment, or waiting for departure exceeds the physical or staffing capacity of the ED (Australasian College for Emergency Medicine 2002). This overcrowding phenomenon has been reported in numerous countries and cities around the globe for decades, e.g., Gallagher and Lynn (1990), Cowan and Trzeciak (2005), Derlet et al. (2001), Miro et al. (1999), Shih et al. (1999), and is observed to be worsening in recent years (Hing and Bhuiya 2012; Horwitz and Bradley 2009). This pervasive problem has drawn a lot of attention from the public because it can lead to many grievous consequences such as public safety at risk, prolonged pain and suffering, and long waits and dissatisfaction of patients (Derlet and Richards 2000). To better understand the ED overcrowding phenomenon, hospital administrators and medical researchers have studied various causes such as increased complexity and acuity of patients presenting to EDs,

Y.-H. Kuo (✉)
Stanley Ho Big Data Decision Analytics Research Centre, Chinese University of Hong Kong, Shatin, New Territories, Hong Kong, China
e-mail: yhkuo@cuhk.edu.hk

J.M.Y. Leung
Department of Systems Engineering and Engineering Management, Chinese University of Hong Kong, Shatin, New Territories, Hong Kong, China
e-mail: janny@se.cuhk.edu.hk

C.A. Graham
Accident and Emergency Medicine Academic Unit, Chinese University of Hong Kong, Shatin, New Territories, Hong Kong, China
e-mail: cagraham@cuhk.edu.hk

A. Matta et al. (eds.), *Health Care Systems Engineering for Scientists and Practitioners*, Springer Proceedings in Mathematics & Statistics 169,
DOI 10.1007/978-3-319-35132-2_16

overall increase in patient volume, managed care problems and lack of beds for patients admitted to hospitals (Derlet and Richards 2000). There is also a huge literature on ED overcrowding; we refer the reader to Hoot and Aronsky (2008), Hwang et al. (2011), Richardson and Hwang (2001). Most of the studies suggest that there is a mismatch between the demand (i.e., patients) and the supply (i.e., medical resources) for emergency care. While this mismatch, without doubt, is a primary factor for ED overcrowding, it appears that this demand-supply imbalance problem is insoluble because most hospitals are not allowed to reject patients, unless in very exceptional circumstances, and do not have sufficient financial budgets to expand their resources. Thus, with the demand and supply being not adjustable, one alternative to relieve the ED overcrowding situation is to reduce operational inefficiency, which is mainly a system issue (American College of Emergency Physicians 2004). In this paper, we examine the impacts of an example of system problems—missing patients—on ED operational efficiency by conducting a real case and a simulation study, and provide some suggestions on potential solutions to reduce the negative impacts.

We conduct a case study of the ED of the Prince of Wales Hospital (PWH) in Hong Kong and built a simulation model to represent its daily operations. PWH is one of the largest public general hospitals in Hong Kong and also the teaching hospital for the Faculty of Medicine of the Chinese University of Hong Kong. It has around 1500 hospital beds and 4500 staff, and serves the region of New Territories East (of more than 1.5 million people). The ED of PWH, which provides 24-h Accident and Emergency (A&E) services and handles around 430 cases per day, is also facing the challenge of overcrowding. Due to the high quality of emergency care and the very low service charge (HKD 100, around EUR 11.5, per attendance), there is often misuse of the ED by non-urgent patients (Chung 2000). All of these together with the very serious understaffing situation have led to the ED overcrowding problem (Rainer 2013). As a result, non-critical patients may need to experience a very long waiting time of several hours. In preparation for the growing (and aging) population in Hong Kong, the ED was relocated in October 2010 to accommodate the increasing patient demands. One of the major changes after the relocation was to establish a new sub-waiting area for consultation, which is closer to the consultation rooms than before. The primary purpose of having this sub-waiting area is to shorten the walking time of patients to consultation rooms. We also found that this sub-waiting area contributes to reduce "missing" patients: it enables the nurses to more easily notify patients that they will soon be seen by a doctor, so that they would not leave the waiting area (e.g., for a meal or even going back to their living places) while waiting. Consequently, this reduces the inactivity times of doctors waiting for those "missing" patients, and hence reduces the waiting times of subsequent seen patients.

This paper is organized as follows. In the next section, we give a brief literature review on the related work. In section "Real Case Study: Emergency Department of the Prince of Wales Hospital in Hong Kong", we will conduct a case study of the impacts of this change on the ED performance based on the real data collected. We also conduct a simulation study in section "Simulation Study: Impacts of Missing

Patients" to examine how different levels of missing patients and various "missing time" influence the patient waiting time. Section "Conclusions" summarizes our work.

Literature Review

There has been a large literature on the use of operations research techniques for analyzing and improving healthcare systems. For recent literature reviews, we refer the reader to Hulshof et al. (2012), Rais and Viana (2011) and Turner et al. (2010). Here we focus on the applications for the ED systems.

Green et al. (2008) adopted a queueing theory approach to determine ED staffing level and reduce the proportion of patients who leave without being seen. Cochran and Roche (Cochran and Roche 2009) developed a spreadsheet-based tool to size ED areas. Their tool was based on a multi-class queueing network and captured different hospital-specific differences such as patient acuity mix, arrival patterns, and efficiencies of different processes. Mayhew and Smith (2008) used a queueing model to evaluate completion times in EDs, analyze the UK Government's 4-h completion time target (i.e., a target that 90 % of completions should be within 4 h), and assess the feasibility of future A&E targets. Saghafian et al. (2012) proposed a combination of analytic and simulation models to examine the effectiveness of patient streaming policies. While these papers demonstrated that analytical models can be applied to analyze ED operations and enhance system performances, we adopt a simulation approach to model the ED system because it can incorporate randomness and many complicating components, such as time and category-dependent arrival rates of patients, different service-time distributions, and time-varying staffing levels, into the model. A simulation model also enables us to freely examine different "what-if" scenarios and easily explain the ideas to practitioners, who are in general not equipped with advanced knowledge in operations research.

The applications of system simulation in the area of health-care management have been studied for more than half of a century, e.g., Fetter and Thompson (1965); and the academic literature on simulation in health-care is immense. We refer the reader to Brailsford et al. (2009), Günal and Pidd (2010), Jahangirian et al. (2012), Jun et al. (1999) for an overview. Hoot et al. (2008) developed a discrete event simulation tool, called ForecastED, to predict near-future ED operational measures with the current patient information. Rossetti et al. (1999) developed a simulation tool to aid physician scheduling, by examining different work schedules and evaluating their impacts on ED performance such as patient throughput and resource utilization. Ahmed and Alkhamis (2009) proposed a simulation-optimization approach to find the optimal numbers of medical staff required, including physicians, lab technicians, and nurses, to maximize patient throughput and to reduce patient time in the system. Connelly and Bair (2004) developed a simulation model for a system-level investigation of ED processes.

By using the simulation model, they compared a fast-track triage approach with an acuity ratio triage approach. Lane et al. (2000) used simulation to analyze the functioning of the ED system and the policies with different bed capacity and demand pattern scenarios. Baesler et al. (2003) used simulation to estimate patients' time in system and the maximum level of patients' demand that the ED can absorb.

In the literature, missing patients in EDs are rarely addressed. Although there is research on patients who leave without being seen, these patients are only an indicator that patients are dissatisfied with the ED performance, but will not worsen the overcrowding situation (and may even make the ED more efficient as they have left the system and reduced workload from physicians). On the other hand, missing patients will cause operational inefficiency because there is physician inactivity time while waiting for patients and the missing patients will come back for consultation some time later (and so the physician workload does not change). The missing patients phenomenon is more common in densely populated cities as non-urgent patients may choose to wait at home, which is within a short distance, when they realize that they may have to wait for many hours. In the next section, we will present a real case study of an ED in Hong Kong.

Real Case Study: Emergency Department of the Prince of Wales Hospital in Hong Kong

The ED of PWH handles around 430 cases per day. There are five categories of patients: 1(critical), 2(emergency), 3(urgent), 4(standard) and 5(non-urgent). The lower the category, the more urgent the patient is. Categories 1 and 2 patients have life-threatening conditions and must be treated immediately after admitted to the ED; they will be rushed to resuscitation rooms directly, without registration and triage, and will be provided timely medical treatments. Other patients (i.e., categories 3–5 patients), after registration, are assessed by a triage nurse and classified by category so as to assign priorities for receiving treatments. In our work and the rest of this paper, we put category 5 patients into category 4 as they have the same flow and priority in real practice and there are only small proportion of category 5 patients. Categories 1 and 2 patients have the highest priority and category 3 patients have a higher priority over category 4 patients. Within the same category, patients are seen on a first-come, first-served (FCFS) basis. In addition to the triage categories, patients are also classified by their mobility. In the day time, mobile patients, called walking patients, are treated in the Walking division, while patients on a trolley-bed or a wheel-chair, called non-walking patients, are treated in the Non-walking division. In the nighttime, in order to better utilize the reduced workforce, all patients will be treated in the same division.

In October 2010, the ED of PWH was relocated to a new building with a new layout. One of the major changes in the layout and the patient flow is that a sub-waiting area for consultation was introduced; when patients are soon to be seen by a physician, they will be called to move from the main waiting area to the sub-waiting area. This enables the nurses to easily notify the patients that they will soon have their consultation and should not leave the sub-waiting area. The nurses can also realize quickly when patients who will soon see physicians are currently not around.

To analyze the effectiveness of this relocation, we compare the net times from triage to consultation for category 4 patients, who are mostly walking patients, before and after the relocation. To make fair comparisons, we present the data of the month of December 2009 (when operating in the old location) and December 2010 (after relocation). (We chose to report on the second month after the relocation because there was a "warm-up" period since staff members needed time to get familiar with the new layout and settings.) In these two periods, the staffing levels were the same. There were 12,945 patient visits in December 2009, and 13,287 visits in December 2010, which translate to around 418 and 429 cases per day respectively. Although there was an increase of 2.64 % in the total number of patient visits, the average net time from triage to consultation for category 4 patients decreased from 112.91 min to 107.77 min (a 4.55 % decrease). Figure 1 also shows that after the relocation, more patients could be seen within 30, 120, 150, 180 and 210 min after triage. The improvement in service is particularly notable for patient waits between 2–3.5 h. 83.99 % of category 4 patients were seen within 3 h in the new layout while only 78.83 % could be seen within 3 h in the old layout. To test if the claim is statistically significant, we conduct a two-sample t-test to examine if the average net time from triage to consultation for category 4 patients decreased from 2009 to 2010.

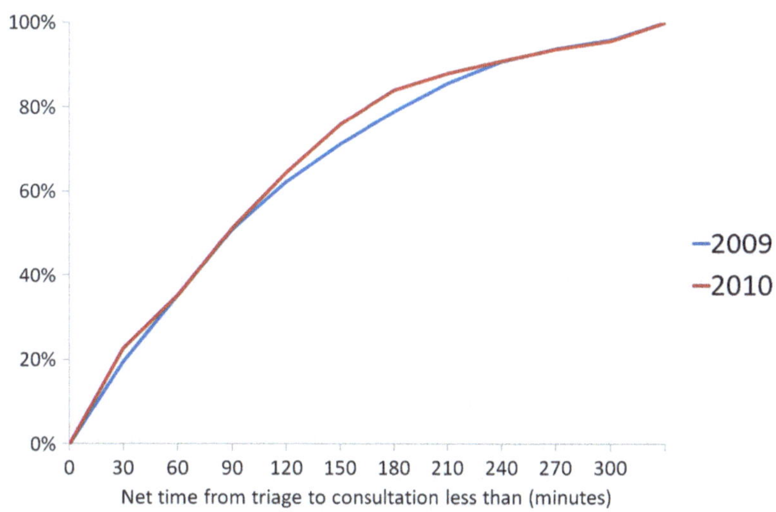

Fig. 1 Net time from triage to consultation for category 4 patients

H_0 : there is no difference between the average net times from triage to consultation for category 4 patients in 2009 and 2010.
H_1 : the average net time from triage to consultation for category 4 patients in 2010 is less than the one in 2009.

The t-statistic is 3.31 and the p-value is 0.001. Hence, the null hypothesis is rejected at the levels greater than 0.001. All the above suggest that patients benefited from this relocation and the change in layout.

Simulation Study: Impacts of Missing Patients

Based on the system of the ED of PWH, we developed a detailed simulation model using the simulation software ARENA to capture the processes and the flow of the ED. Our simulation model captures: all relevant treatment processes (triage, consultation, lab tests, etc.), the complexities of intertwining and re-entrant patient-flows, complicated arrival rates that vary by time and patient category and adjustable staff deployment (shift, breaks, doctors on reserve). The necessary input parameters/data are arrival rates, probability distributions of service times, available resources and schedules of doctors and nurses. In the simulation model, patients are generated according to category-dependent non-homogeneous Poisson processes. Upon arrival, a patient has to register at the registration counter and then be assessed by a triage nurse to assign a triage category. If the registration staff or the triage nurse is busy, the patient will have to wait in line and be seen on a FCFS basis. After triage, the patient will need to wait for a physician consultation. Patients are seen according to their triage category; within the same triage category, they will be seen on a FCFS basis. After consultation, patients may be required to undergo medical diagnostic examinations (e.g., X-ray, blood test) and revisit the doctor after their medical reports are ready. They can be discharged from the ED after all the required treatments and medical diagnostic examinations are done. For registration, triage, and consultation, walking and non-walking patients are treated separately in two different divisions in the daytime, but will be treated together in the nighttime. The parameters of our simulation model were estimated from the data collected by the ED, where parameters for service-time distributions were indirectly estimated with a simulation-optimization approach. For more details about our simulation model, our parameter estimation procedure, and model validation, we refer the reader to Kuo et al. (2016), Rado et al. (2014).

We use the simulation model with the data provided by the ED, including patient arrivals and staffing levels, to analyze the impacts of missing patients on ED performance. We examine scenarios of different percentages of missing patients and durations of missing time. Since non-walking patients and category 3 or below patients are unlikely to leave while waiting, category 4 walking patients are the only type that have missing patients. We consider three levels of missing patients: 5 %, 10 %, and 15 % of category 4 walking patients are missing. Since category

4 walking patients are about half of the total patients in the ED, the three levels of missing patients are respectively around 2.5 %, 5 %, and 7.5 % of the total number of patients. We model the missing time by an exponential random variable and consider three average missing durations: 20, 40, and 60 s.

We conduct a full-factorial design of experiments and compare the net times from registration to consultation, which reflect the waiting times for consultation upon arrival for category 4 walking patients. For each scenario, we ran 100 replications of simulation runs of 34 days, where each simulation replication started from an empty system. The first 3 days was a warm-up period and not included in our reported statistics. In other words, we simulated patient flows of 100 independent months, each starting from a non-empty system. We calculatc the averages of category 4 walking patients' waiting times for consultation upon arrival among the 100 replications, as reported in Table 1. We observe that although the missing times are very small (less than 1 %) compared to the waiting times, their impacts on ED performance are notable, where the increase in waiting time for category 4 walking patients ranges from 5.43 % to 26.70 %. This indicates that missing patients can induce significant operational inefficiency. If EDs can reduce the number of missing patients and their missing time, for example by redesigning layout as presented in section "Real Case Study: Emergency Department of the Prince of Wales Hospital in Hong Kong" or advanced technologies to be discussed in the next section, a remarkable reduction in patient waiting time can be achieved. Figure 2 presents the box plot for waiting time for category 4 walking patients. As expected, the variation in waiting times appear to be larger when the proportion of missing patients is higher or the missing time is longer.

Table 1 Net times (in minutes) from registration to consultation for category 4 walking patients

Instance	Percentage of missing patients (%)	Average missing time (s)	Waiting time (min)	Percentage increase (%)
1	0	0	176.64 (2.00[a])	0.00[b]
2	5	20	186.24 (2.32)	5.43
3	5	40	190.86 (3.05)	8.05
4	5	60	197.97 (3.08)	12.08
5	10	20	193.05 (2.80)	9.29
6	10	40	200.36 (3.12)	13.43
7	10	60	213.66 (3.59)	20.96
8	15	20	194.48 (2.98)	10.10
9	15	40	212.93 (3.91)	20.54
10	15	60	223.80 (3.89)	26.70

[a]Indicates the half width

[b]Indicates the percentage increase in the waiting time compared with instance 1

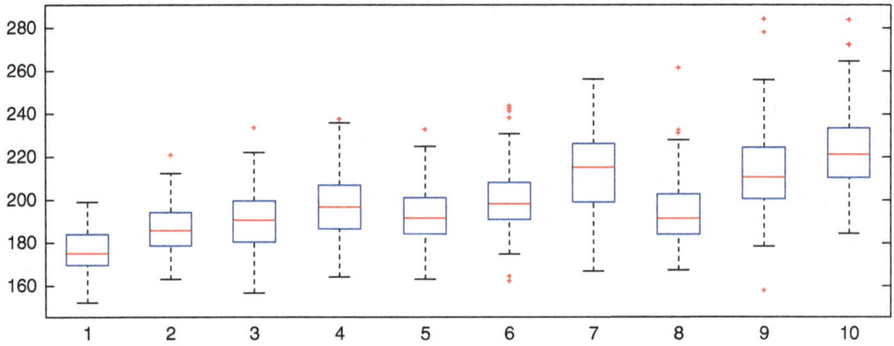

Fig. 2 Net time from registration to consultation for category 4 walking patients

Potential Solutions for Reducing Missing Patients

In sections "Real Case Study: Emergency Department of the Prince of Wales Hospital in Hong Kong" and "Simulation Study: Impacts of Missing Patients", we showed that missing patients can induce significant operational inefficiency. Therefore, administrators of EDs should work out solutions for reducing missing patients and thereby improve ED performance. A potential solution for this problem is to adopt information technology to reduce the number of missing patients in the system.

One of the main reasons for having missing patients is that patients may leave EDs when they realize that they need to wait for many hours. Some of the patients may choose to wait at other places (e.g., home) outside the ED and may miss the call for consultation. A solution for this problem is to provide patients with the up-to-the-minute information about their waiting status even when they are out of the ED. For example, in order to alleviate and improve the overcrowding problem at pharmacy, United Christian Hospital in Hong Kong developed a mobile app called DREAMS (Drugs Ready E-Alert Me System) to notify patients that their drugs are ready for collection so that it is not necessary for patients to be waiting at pharmacy, as the waiting time is usually quite long (United Christian Hospital 2014). Similarly, EDs can also consider to develop mobile apps to notify patients that they will soon be seen by a doctor and should stay in the designated area for waiting for consultation.

Another solution is to identify the locations of patients. There has been much research on the implementation of radio-frequency identification (RFID) in EDs, e.g., Huang et al. (2010), Pérez et al. (2012). In an RFID-enabled environment, each patient will be given a RFID tag at the time he/she registers. As a patient moves around the ED, the RFID system can keep track of the locations and the times that this patient passes through. This enables the nurses to locate patients and realize if a patient has left the ED. Furthermore, in EDs, RFID technology can also improve patient safety, e.g., monitoring critical quality metrics and ensuring correct medication.

Conclusions

ED overcrowding has been reported over many years around the globe and is known to have numerous negative patient health outcomes. One way to alleviate the overcrowding phenomenon is to enhance operational efficiency. In this paper, we examine the impacts of missing patients on ED performance by conducting a real case study and a simulation study. Both of the studies show that missing patients can induce operational inefficiency in EDs. Simulation results show that even a small proportion of patients missing and a short missing time can increase patient waiting time significantly. We suggest that EDs should consider to adopt information technology to reduce the operational inefficiency due to missing patients.

Acknowledgements The research of the first author is supported by Microsoft Research Asia Collaborative Research Fund FY15-RES-THEME-049 and Macao Science and Technology Development Fund 088/2013/A3. The research of the second author is partially supported by GRF grant 414313 from the Hong Kong Research Grant Council. The authors would also like to thank Mr. Stones Wong, Operations Manager of the Emergency Department of the Prince of Wales Hospital, for his assistance in data collection.

References

Ahmed, M.A., Alkhamis, T.M.: Simulation optimization for an emergency department healthcare unit in Kuwait. Eur. J. Oper. Res. **198**(3), 936–942 (2009)

American College of Emergency Physicians. Efficiency in the Emergency Department: Doing things faster without sacrificing quality. ACEP Reference and Resource Guide (2004)

Australasian College for Emergency Medicine: Policy document — standard terminology. Emerg. Med. **14**, 337–340 (2002)

Baesler, F.F., Jahnsen, H.E., DaCosta, M.: The use of simulation and design of experiments for estimating maximum capacity in an emergency room. In: Proceedings of the 2003 Winter Simulation Conference, pp. 1903–1906 (2003)

Brailsford, S.C., Harper, P.R., Patel, B., Pitt, M.: Analysis of the academic literature on simulation and modelling in health care. J. Simul. **3**, 130–140 (2009)

Chung, C.H.: Emergency department misuse and administrative interventions. Hong Kong J. Emerg. Med. **7**(4), 220–229 (2000)

Cochran, J.K., Roche, K.T.: A multi-class queuing network analysis methodology for improving hospital emergency department performance. Comput. Oper. Res. **36**(5), 1497–1512 (2009)

Connelly, L.G., Bair, A.E.: Discrete event simulation of emergency department activity: a platform for system-level operations research. Acad. Emerg. Med. **11**(11), 1177–1185 (2004)

Cowan, R.M., Trzeciak, S.: Clinical review: emergency department overcrowding and the potential impact on the critically ill. Crit. Care **9**(3), 291–295 (2005)

Derlet, R.W., Richards, J.R.: Overcrowding in the nation's emergency departments: complex causes and disturbing effects. Ann. Emerg. Med. **35**(1), 63–68 (2000)

Derlet, R.W., Richards, J.R., Kravitz, R.L.: Frequent overcrowding in US emergency departments. Acad. Emerg. Med. **8**(2), 151–155 (2001)

Fetter, R.B., Thompson, J.D.: The simulation of hospital systems. Oper. Res. **13**(5), 689–711 (1965)

Gallagher, E.J., Lynn, S.G.: The etiology of medical gridlock: causes of emergency department overcrowding in New York City. J. Emerg. Med. **8**(6), 785–790 (1990)

Günal, M.M., Pidd, M.: Discrete event simulation for performance modelling in health care: a review of the literature. J. Simul. **4**(1), 42–51 (2010)

Green, L.V., Soares, J., Giglio, J.F., Green, R.A.: Using queueing theory to increase the effectiveness of emergency department provider staffing. Acad. Emerg. Med. **13**(1), 61–68 (2008)

Hing, E., Bhuiya, F.A.: Wait Time for Treatment in Hospital Emergency Departments, 2009. US Department of Health and Human Services, Centers for Disease Control and Prevention, National Center for Health Statistics (2012)

Hoot, N.R., Aronsky, D.: Systematic review of emergency department crowding: causes, effects, and solutions. Ann. Emerg. Med. **52**(2), 126–136 (2008)

Hoot, N.R., LeBlanc, L.J., Jones, I., Levin, S.R., Zhou, C., Gadd, C.S., Aronsky, D.: Forecasting emergency department crowding: a discrete event simulation. Ann. Emerg. Med. **52(5)**, 116–125 (2008)

Horwitz, L.I., Bradley, E.H.: Percentage of US emergency department patients seen within the recommended triage time: 1997 to 2006. Arch. Intern. Med. **169**(20), 1857–1865 (2009)

Huang, Y.C., Chu, C.P., Lin, Y.S., Kuo, C.H.: RFID Applications in Hospitals—A Case Study for Emergency Department (2010)

Hulshof, P.J., Kortbeek, N., Boucherie, R.J., Hans, E.W., Bakker, P.J.: Taxonomic classification of planning decisions in health care: a structured review of the state of the art in OR/MS. Health Syst. **1**(2), 129–175 (2012)

Hwang, U., McCarthy, M.L., Aronsky, D., Asplin, B., Crane, P.W., Craven, C.K., Epstein, S.K., Fee, C., Handal, D.A., Pines, J.M., Rathlev, N.K., Schafermeyer, R.W., Zwemer, F.L., Bernstein, S.L.: Measures of crowding in the emergency department: a systematic review. Acad. Emerg. Med. **18**(5), 527–538 (2011)

Jahangirian, M., Naseer, A., Stergioulas, L., Young, T., Eldabi, T., Brailsford, S., Patel, B., Harper, P.: Simulation in health-care: lessons from other sectors. Oper. Res. **12**, 45–55 (2012)

Jun, J.B., Jacobson, S.H., Swisher, J.R.: Application of discrete-event simulation in health care clinics: a survey. J. Oper. Res. Soc. **50**(2), 109–123 (1999)

Kuo, Y.H., Rado, O., Lupia, B., Leung, J.M., Graham, C.A.: Improving the efficiency of a hospital emergency department: a simulation study with indirectly imputed service-time distributions. Flex. Serv. Manuf. J. **28**(1), 120–147 (2016)

Lane, D.C., Monefeldt, C., Rosenhead, J.V.: Looking in the wrong place for healthcare improvements: a system dynamics study of an accident and emergency department. J. Oper. Res. Soc. **51**(5), 518–531 (2000)

Mayhew, L., Smith, D.: Using queuing theory to analyse the Government's 4-h completion time target in accident and emergency departments. Health Care Manag. Sci. **11**(1), 11–21 (2008)

Miro, O., Antonio, M.T., Jimenez, S., De Dios, A., Sanchez, M., Borras, A., Milla, J.: Decreased health care quality associated with emergency department overcrowding. Eur. J. Emerg. Med. **6**(2), 105–107 (1999)

Pérez, M.M., Cabrero-Canosa, M., Hermida, J.V., García, L.C., Gómez, D.L., González, G.V., Herranz, I.M.: Application of RFID technology in patient tracking and medication traceability in emergency care. J. Med. Syst. **36**(6), 3983–3993 (2012)

Rado, O., Lupia, B., Leung, J.M., Kuo, Y.H., Graham, C.A.: Using simulation to analyze patient flows in a hospital emergency department in Hong Kong. In: Proceedings of the International Conference on Health Care Systems Engineering, pp. 289–301 (2014)

Rainer, T.: Why are Hong Kong's emergency departments so understaffed? Emerg. Physicians Int. **11**, 26–27 (2013)

Rais, A., Viana, A.: Operations research in healthcare: a survey. Int. Trans. Oper. Res. **18**(1), 1–31 (2011)

Richardson, L.D., Hwang, U.: Access to care a review of the emergency medicine literature. Acad. Emerg. Med. **8**(11), 1030–1036 (2001)

Rossetti, M.D., Trzcinski, G.F., Syverud, S.A. : Emergency department simulation and determination of optimal attending physician staffing schedules. In: Proceedings of the 1999 Winter Simulation Conference, pp. 1532–1240 (1999)

Saghafian, S., Hopp, W.J., Van Oyen, M.P., Desmond, J.S., Kronick, S.L.: Patient streaming as a mechanism for improving responsiveness in emergency departments. Oper. Res. **60**(5), 1080–1097 (2012)

Shih, F.Y., Huel-Ming, M., Chen, S.C., Wang, H.P., Fang, C.C., Shyu, R.S., Huang, G.T., Wang, S.M.: ED overcrowding in Taiwan: facts and strategies. Am. J. Emerg. Med. **17**(2), 198–202 (1999)

Turner, J., Mehrotra, S., Daskin, M.S.: Perspectives on health-care resource management problems. In: A Long View of Research and Practice in Operations Research and Management Science, pp. 231–247 (2010)

United Christian Hospital (2014): DREAMS. http://www3.ha.org.hk/uch/hospital/show.asp?c=spIQDS Accessed 16 February 2015

System Dynamics Modelling of Emergent and Elective Patient Flows

Paolo Landa, Michele Sonnessa, Elena Tànfani, and Angela Testi

Introduction

Emergency Department (ED) overcrowding together with waiting lists for elective surgery are the two most substantial issues in publicly funded health systems. They are regularly reported in newspaper headlines; it seems that health resources are not organized in a manner consistent with the true needs of citizens. Moreover, citizen's needs are in competition with each other because elective patients, who are entitled to an intervention, must delay their intervention due to unpredictable excessive arrivals in the ED that require a hospital admission. Why is this situation not predicted, as it happens regularly following a seasonal pattern? What is the best and most affordable remedy? These issues are addressed in this paper.

In recent years, two approaches to address ED overcrowding have been suggested. The first is intended to reduce the number of patients using the ED inappropriately, and the second is directed at facilitating early discharges from inpatient wards to face the increase of the emergent admissions during peak periods (Bagust et al. 1999). The problem related to emergent patients who could safely be treated in more appropriate care settings is well known in almost all countries. However, these improper accesses (in particular "white" urgency codes), which

P. Landa (✉)
Department of Economics and Business Studies, University of Genova, Genova, Italy

University of Exeter Medical School, Exeter (United Kingdom)
e-mail: paolo.landa@unige.it; P.Landa@exeter.ac.uk

M. Sonnessa • E. Tànfani • A. Testi
Department of Economics and Business Studies, University of Genova, Genova, Italy

A. Matta et al. (eds.), *Health Care Systems Engineering for Scientists and Practitioners*, Springer Proceedings in Mathematics & Statistics 169,
DOI 10.1007/978-3-319-35132-2_17

constitute approximately 30 % of the total accesses in Italy (Agenas 1999), are managed quickly and take less than 15 % of the total hours of the ED staff; they often comprise a social rather than a health emergency.

Instead, the largest commitment of ED personnel is on more serious patients (green, yellow and red urgency codes) who must be hospitalized within 6 h according to the current reference national standard (Simeu 2011). The crucial issue pertains to patients staying on trolleys, waiting to be admitted to a hospital ward. It is therefore necessary to better manage the second approach (i.e., designing interventions focused on smoothing intra-hospital patient flows from the ED to inpatient wards and accelerating discharges towards low intensity care and home care settings). Addressing ED overcrowding involves recognizing that it is necessary to increase bed capacity to accommodate emergent patients. This approach, however, is very difficult to implement because the number of hospital beds is currently decreasing due to financial concerns of public expenditure. Particularly, Italy has the lowest ratio of beds to population amongst the countries of Europe (approximately 3.4 per 1000 inhabitants vs. 6.3 European average, Eurostat). In this context, increasing the ED capacity without considering that most patients must be transferred to inpatient wards can worsen the problem (IHI 2003).

The specific objective of this paper is to develop a decision support tool to be used by decision makers and politicians to simultaneously analyse competing patient flows and to fine tune the system. The tool can be used to determine the best strategy for addressing excessive ED crowding without increasing waiting lists for elective patients. The developed tool is based on a System Dynamics (SD) model that is able to predict how to allocate the entire hospital bed capacity to facilitate the flow of patients across the system. The focus of the model is, therefore, patient flow analysis, instead of individual patient pathways inside the ED and the inpatient wards. For this reason, SD simulation has been chosen as the best modelling technique to understand the entire system and the relationships between the various system components (Brailsford et al. 2004).

Generally, the major advantage of using a simulation for the analysis and evaluation of health services is the ability to perform an analysis of "what if", which evaluates the performance of the system in various scenarios by varying the input data and model parameters and identifying critical points and system bottlenecks. Two recent literature reviews (Günal and Pidd 2010; Abouljinane et al. 2013) indicate that a major field of application of simulation concerns the performance analysis in Emergency Departments. Amongst the articles that show real applications, most address the use of Discrete Event Simulation (DES), with the aim of improving the flow of patients and increase throughput (Abo-Hamad and Arisha 2013; Paul and Lin 2012; Proudlove et al. 2007). Most of the studies present the implementation of simulation models starting from case studies of specific hospitals (Duguay and Chetouane 2007; Katsaliaki and Mustafee 2010). Some authors proposed the use of system dynamics simulation to assist in solving the main strategic problems arising in the ED and to help clinicians and hospital management in reduce the delays and discomforts experienced by patients in the emergency department (Brailsford et al. 2004; Lane et al. 2000; Vanderby and Carter 2010; Wong et al. 2012). The main advantage of SD is its ability to be used

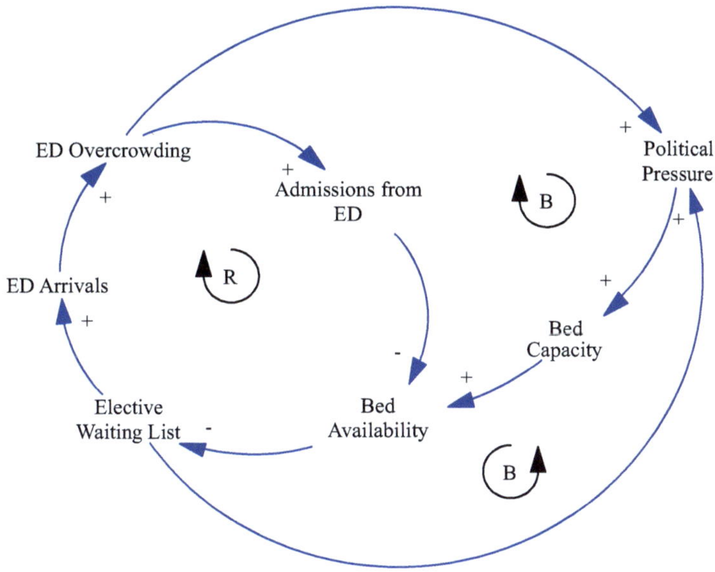

Fig. 1 Causal loop diagram of the system

at a higher, more aggregated and strategic level than DES. Additionally, the data requirements for developing a SD model are much less than for DES (Brailsford et al. 2014).

Casual Loop Diagram

According to Brailsford (2008), in SD models, the concepts of feedback and causal effects are important and analysed by combining qualitative and quantitative aspects. In particular, the qualitative aspect involves the construction of a causal loop diagram, which graphically depicts the relationship and interaction amongst the system elements (Fig. 1). In this study, the central concept of the causal loop diagram is related to understanding how the patient flows and system stocks interact through feedback loops in which a change in one variable affects other variables. The diagram shown in Fig. 1 includes qualitative information obtained from a deep observation and analysis of the system and data collection. The loop diagram mainly describes the process by which ED overcrowding affects hospital activities. A first feedback loop, on the left side of Fig. 1, shows that ED overcrowding entails increasing the number of patient admissions in inpatient wards. Of course, not all patients who arrive in the ED need further treatments. Some of them are discharged, but a consistent percentage (approximately 50 % on average in Italian hospitals) must be admitted in an inpatient ward to be further diagnosed or treated. This need means occupying ward beds and leaving less bed capacity available for planned patient admissions, thus increasing the elective waiting lists.

Competition between emergent and elective patients on for beds on the same ward increases dramatically and is often exacerbated by measures aimed to stop elective admissions and transfer patients to other hospitals, thus causing long waits on trolleys for emergent patients before being admitted to inpatient wards. Indeed, when the bed occupancy approaches full occupancy, the elective waiting lists increase due to the effect of blocking elective flows in advance, as in the case study presented later. When the waiting time is excessive, patients in more critical conditions bypass the elective waiting list and go to the ED directly, thus further increasing ED arrivals. This process is partly a direct result of the deterioration in their health and partly due to a behavioural response of doctors wishing to have their patients admitted "by the back door" (Lane et al. 2000). In conclusion, a reinforcing loop is to be expected.

The only way to balance the system is the political pressure set in motion by both ED overcrowding and excessive wait times. Both situations usually receive substantial media coverage in newspapers and are deeply felt by politicians and health managers. Political pressure can, therefore, trigger two loops that are able to balance the system. The decisional variable is the bed capacity that should be increased, by means of various organizational instruments, such as providing additional shifts of personnel to keep some wards open that are usually closed on the weekend or anticipating discharges from acute to low-intensity wards.

Based on the qualitative analysis of the problem presented above, the rest of the paper is organized as follows. In Section "Modelling framework", the description of the SD simulation model structure is given by means of stock and flow diagrams. In Section "Application to a Real Case Study", a case study is introduced together with the main data collected, and in Section "Validation and Results", the scenario analyses and results are reported.

Modelling Framework

The stock-flow model used to represent the system under analysis is composed of two main sub processes, as depicted in Figs. 2 and 3. The first process captures the patient flows within the Emergency Department (ED), whereas the second represents the input-output flows involved in the inpatient hospital wards subsystem.

In the first process, three stocks are modelled representing the queues of *Patients in ED*, *Patients in Short Observation Unit* (SOU) and *Patients to be Admitted in Inpatient Wards* after the ED treatment (Fig. 2).

The SOU represents a buffer that is able to delay some patient flows towards hospital wards. It is designed for patients who require exams to determine the seriousness of their condition or a short course of treatment for conditions that may be rapidly resolved and patients are likely to be discharged within 72 h.

The main model flows are: (1) *Emergent Patient Arrivals* who enter the *ED* stock, where they wait to be processed; (2) *Discharges* (i.e., patients exiting the ED); (3) *Short Observation Unit Admissions*; and (4) *Admissions in Inpatient Wards*, which

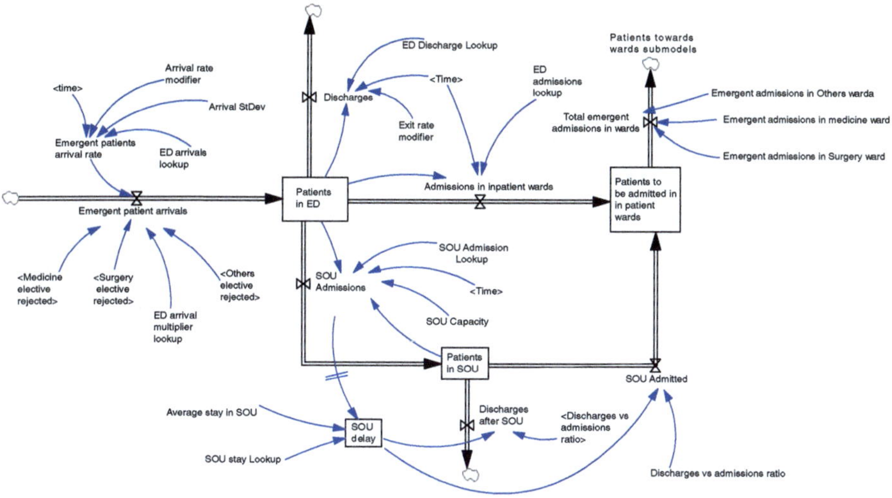

Fig. 2 ED sub-model

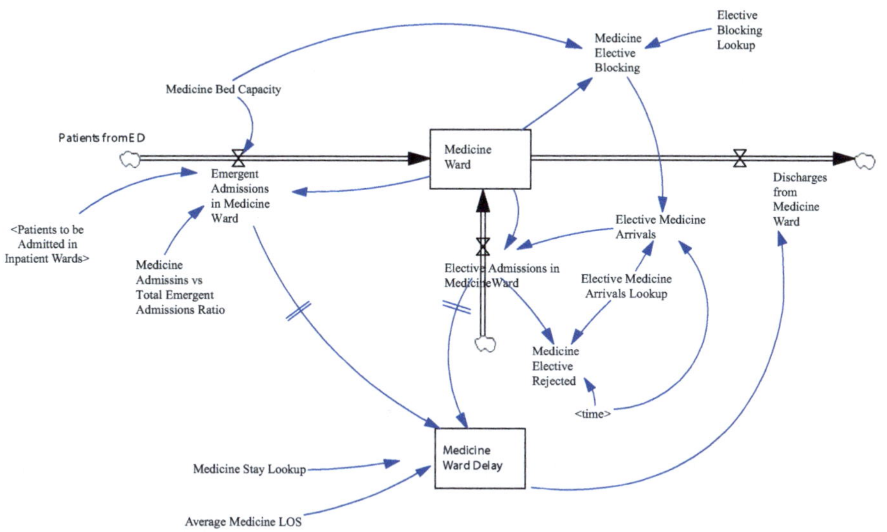

Fig. 3 Medicine ward sub-model

represent patients who must be admitted to an inpatient hospital ward after ED treatment and diagnosis. Note that the flow of patients treated in the SOU unit is further divided into *Discharged after SOU* and *SOU Admitted* flows.

The rate of arrival and exit of ED patients are modelled using empirical lookup functions based on a weekly time schedule. Indeed, a non-linear lookup function (*Arrival Rate Modifier*) was created to modify the arrival rate of patients to ED, giving the model the capability to adapt to external changes and catch exogenous variations in the input flow.

The elective admissions and the discharge flow from the wards, which are specified in the ward sub-models (Fig. 3), mainly affect the consistency of the *Patients to be Admitted in Inpatient Ward* stocks. Indeed, its current state is the difference between the incoming flows from the ED and the SOU and the outflows of patients admitted to a hospital ward. Each ward sub-model calculates the number of emergent admissions interactively by computing the number of patients who need a ward admission for which a bed is available.

The second sub-process refers to the inpatient wards. Various levels of details can be used. We chose to group wards into macro-specialties, in which beds can be considered adequate substitutes.

In Fig. 3, the sub-model of the Medicine ward is depicted. The other groups are modelled in a similar way. The stock of patients in the ward (*Patients in Medicine Ward*) is the result of two incoming flows (*Emergent admissions in Medicine Ward* and *Elective Admissions in Medicine Ward*) and one exit flow (*Discharges from Medicine Ward*).

If the level of bed occupancy in each inpatient ward reaches a given value, two opposite effects are generated by means of two lookup functions aimed at capturing the feedback loop between ED overcrowding and the waiting list for elective admissions, which was emphasized in the causal loop discussion (Fig. 1).

The first function (*Elective Blocking*) introduces a break in the elective arrivals flow, which depends on the value of the occupancy rate. The break entities and the threshold bed occupancy rates are implemented according to a function decided by the hospital doctors. The second lookup function (*ED Arrivals Multiplier*) models the feedback arising between the increase in the number of elective patients waiting to be admitted to hospital and the increase in the patient arrivals to the ED using a multiplier, which reflects what usually happens in real situations. If the elective waiting lists become longer and the waiting time increases, some patients can be pressured to skip the elective waiting list and go to the ED directly, thus increasing the ED arrivals. This process is partly a direct result of the deterioration in their health and is partly a behavioural response of doctors wishing to have their patients admitted "by the back door" (Lane et al. 2000).

Finally, to improve the model fidelity and its ability to mimic the real system, a third lookup function (*Exit Rate Modifier*) was introduced. If the queue of patients to be treated in the ED increases to a priori decided critical values, this function models the increase in the discharge flow rate to limit the exponential increase in the stock of patients in the ED.

This response corresponds to a reaction of hospital managers to manage unpredictable and unmanageable sudden increases in the number of patients who access the ED during critical periods. The most reasonable response that is modelled is an increase in the early discharges of non-critical patients, which in reality corresponds to transferring them to lower intensity emergency facilities.

The three effects above allow the model to mimic, with adherence to what happens in reality, the interactions amongst the various patient flows across ED and inpatient wards and the corresponding stocks.

The exit flow is modelled through a profile delay detailing the average length of stay of patients in the ward. An excessive length of stay mainly determines overcrowding or congestion phenomena in the wards. Accordingly, the model can make assumptions about the "service time", specifying the time spent by patients in the system as a control parameter. Instead, the time spent by patients inside the ED is a combination of "service time" and waiting time due to bounces of incoming flows.

Application to a Real Case Study

The model was applied to a hospital in Genova (Italy) with approximately 400 inpatient beds and 40,000 ED accesses per year. To better understand the dynamics of patient flows and the interactions between ED and Hospital wards, data over a 1 year period (from January to December 2012) were collected and analysed. There were approximately 14,000 inpatient admissions. Most were admitted from the ED (65.61 %), whereas 34.49 % were elective patients (Table 1).

The empirical distribution of the input and output flows from the ED is represented in Fig. 4 using the time frame of 1 week.

Table 1 ED and inpatient ward input flows

ED accesses	39,980
Total number of inpatient ward admissions	14,288
Emergent patients from ED	9375
Elective patients	4913

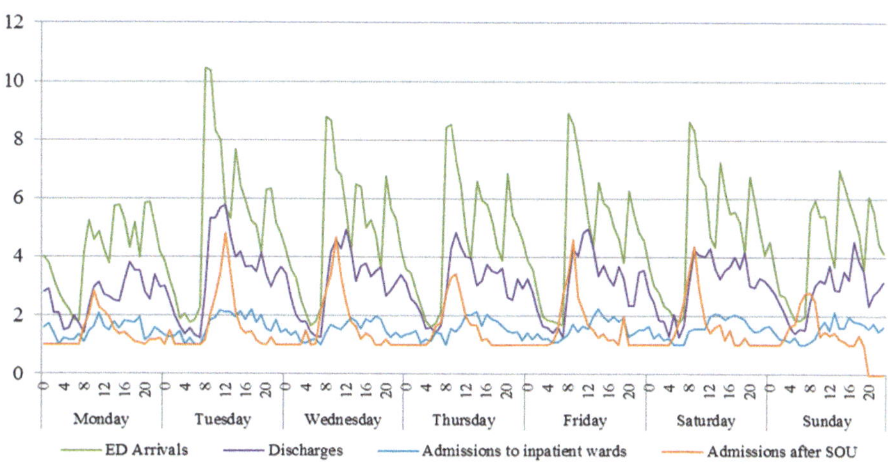

Fig. 4 ED input and output flows

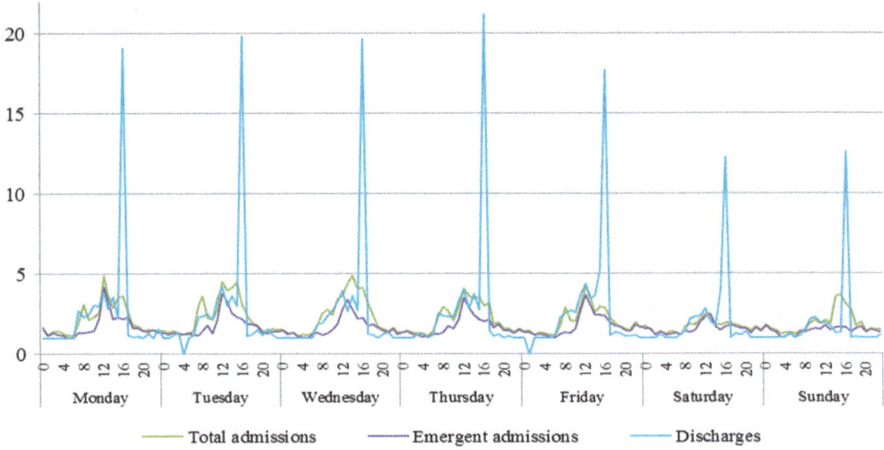

Fig. 5 Inpatient hospital ward input and output flows

The higher peaks in the ED arrivals are usually at approximately 9 a.m. SOU admissions are usually from 2 a.m. until 11 a.m. This peak occurs because in that time period, usually no discharges occur in the hospital, and therefore, SOU is used by the physicians mainly as a buffer for patients waiting to be admitted. Instead, patients are usually admitted to an inpatient ward from 11 a.m. to 5 p.m., when most of the discharges from inpatient wards are performed, thus freeing the stay beds. Finally, the ED discharges are concentrated from 8 a.m. to 8 p.m., with the highest activity of the emergency department, which corresponds to the ordinary daily activities of people.

The input and output flows related to inpatient hospital wards is depicted in Fig. 5. The total input flow includes both the emergent admissions coming from the ED and the planned elective admissions. Admissions to a hospital ward are usually concentrated between 8 a.m. and 1 p.m., whereas the discharges are concentrated from 3 to 5 p.m.

Validation and Results

A face validation was performed during and after the model development. The frame of the model and the main variables and relationships amongst the various system components were designed and verified in collaboration with the doctors and personnel of the hospital and of the Health Regional Agency (ARS Liguria).

Additionally, to justify the model validity, the main output variables of the model were compared with the empirical data that were collected (Tables 2 and 3). A 1 year run was considered and compared to the data collected during 2012.

Table 2 Comparison between simulated output and empirical data

	Empirical data				Simulated data			
Stock and flow	Max	Mean	Median	SD	Max	Mean	Median	SD
Emergent patient arrivals	20	4.55	4	2.99	20.58	4.65	4.13	3.19
Patients in ED	59	24.89	25	7.56	31.05	27.96	26.74	36.65
Patients to be admitted in inpatient wards	27	9.33	9	4.17	14.52	9.81	9.76	1.76
Emergent admissions in other ward	5	0.32	0	0.61	11.16	0.46	0.44	0.19
Emergent admissions in medicine ward	9	0.35	0	0.78	12.28	0.50	0.49	0.20
Emergent admissions in surgery ward	4	0.11	0	0.34	3.8	0.16	0.15	0.62

Table 3 Comparison between simulated output and empirical data

		Empirical data				Simulated data			
Stock	# Beds	Max	Mean	Median	SD	Max	Mean	Median	SD
Medicine ward	150	148	131.46	136	19.03	146.30	135.6	142.10	17.09
Surgery ward	58	56	41.80	42	6.34	53.66	49.39	50.60	5.01
Other ward	168	163	129.85	133	17.86	157.80	142.90	148.70	17.89

In Table 2, in addition to the number of emergent patient arrivals to the ED, another four stock variables are considered (i.e., the number of Patients in the ED, the number of patients admitted to each inpatient ward together with the corresponding total value). For each variable, the max, min, medium, median and standard deviation (SD) per hour are reported. For all variables, acceptable results were obtained.

In Table 3, the same statistics are reported for the average number of patients admitted per hour to the three wards (Medicine, Surgery and Other wards) together with the available bed capacity for each ward (# Beds). The results given in Tables 2 and 3 show that the SD model is able to represent the real system behaviour. Considering several variables of flows and stocks of patients, the overall variation of the mean, max, median and standard deviation and total values is negligible.

The validation results show that the model is able to show how the feedback mechanism amongst various stocks and flows can sustain the system in equilibrium. However, being in equilibrium does not mean that the situation is satisfactory.

The main relevance of the model lies in its ability to describe at what price this equilibrium is obtained (i.e., the potential indicator of dissatisfaction, which in turns triggers the "political pressure" of the causal loop depicted in Fig. 1). This cost mainly consists of organizational changes in the system that are able to improve the overall performance. In this paper, we consider the increase of the inpatient bed capacity, which can be viewed as a major weak point of the system.

In the following, we show how the model can help decision makers by using the substantial increase in ED arrivals due to the winter season flu epidemics as an

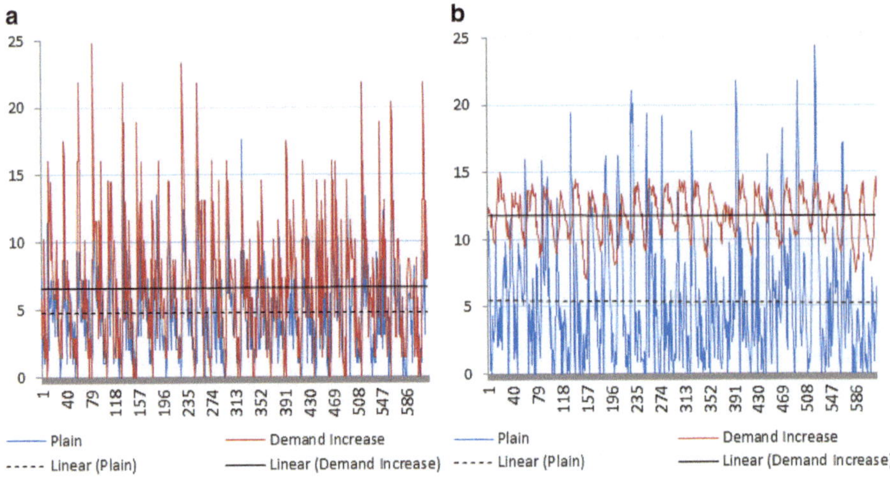

Fig. 6 Effects of a demand increase (**a**) over ED occupancy (**b**)

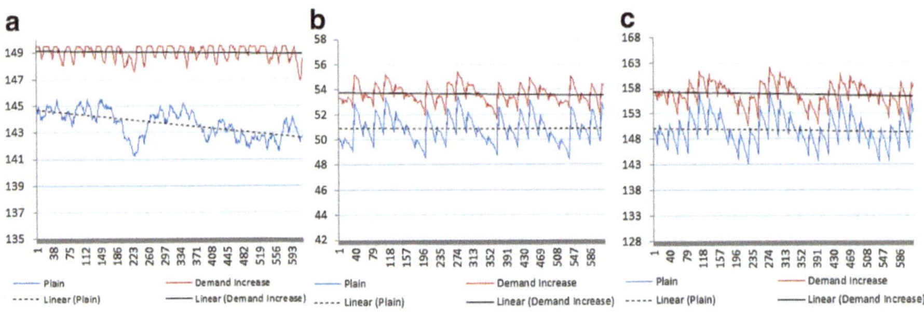

Fig. 7 Bed occupancy in medicine (**a**), surgery (**b**) and other (**c**) wards

example. The increase in patient arrivals to the ED is particularly consistent (60 %) during epidemic flu periods, which has been simulated. The situation is described in Figs. 6, 7 and 8, which depict the model dynamics during approximately 1 month of steady state simulation. The increase in ED arrivals (1) determines a chain of effects that lead to a more than proportional increase in the average number of Patients in the ED (2). Moreover, the increase in ED accesses generates a queue of patients inside the ED that the ED staff is not able to face. Accordingly, the equilibrium of the system is obtained by increasing the early discharges of low-risk patients in the ED.

A greater flow of patients to be admitted and coming from the ED causes a reduction of the available ward beds (Fig. 7), with a significant block of elective patient arrivals (Fig. 8). The block of elective arrivals entails an increase in ED

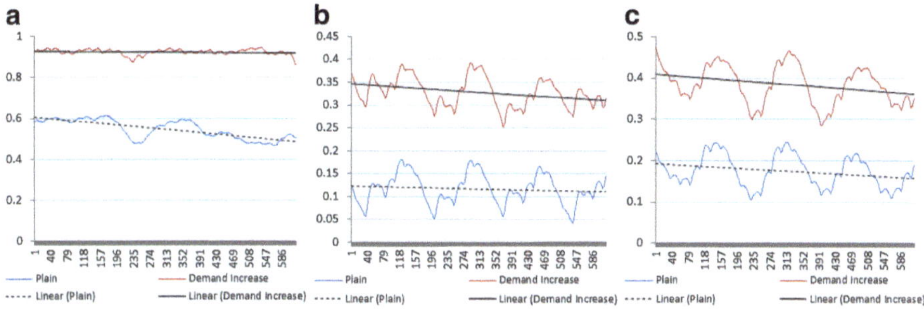

Fig. 8 Elective blocking in medicine (**a**), surgery (**b**) and other (**c**) wards

accesses by elective patients who are not admitted by the hospital. Note that the number of patients in the ED (Fig. 6b) considers both the increase in flu arrivals as well as the increase induced by elective blocking.

The cumulative effect raises stress on the inpatient wards, particularly the Medicine wards, which are mainly used to manage emergent patients and further diagnose them after ED treatment.

Figure 7 shows that the number of occupied beds approaches the total number of beds for each ward (150, 58, 168, respectively, for Medicine, Surgery and Other Wards); specifically, the utilization rate of ward beds increases to 99 % of the overall capacity. This pressure could only be remedied with a bed capacity increase, which seems to be the main bottleneck of the system.

In the absence of such an increase (which in the model is activated by the political pressure loop), the system can be balanced by blocking the elective patient flow.

Figure 8 shows the brake in the elective admission flow caused by the increase in the patients in the ED. The elective blocking in the Medicine ward is more than 90 % but is approximately 35 % and 40 % for surgery and other wards, respectively.

If the waiting time on the elective waiting lists is excessive, patients in more critical conditions bypass the elective waiting list and go to the ED directly, thus further increasing ED arrivals.

The model can thus be used to compute how many beds would be necessary to avoid this difficult situation, leading to a more sustainable equilibrium of the overall (ED and inpatient wards) system. In this particular example, an increase of 25 beds in Medicine wards (from 150 to 175) would stabilize the bed occupation rate and the waiting lists.

Figure 9b shows the effect of the bed capacity increase in returning the elective blocking rate near the level of the base scenario. Note that the dynamics of the Medicine bed occupancy rate (1) are more variable, as it was before the ED demand increased, whereas the average level is quite similar. This result confirms that if the system is under pressure, the bed occupancy rate is almost near full capacity usage.

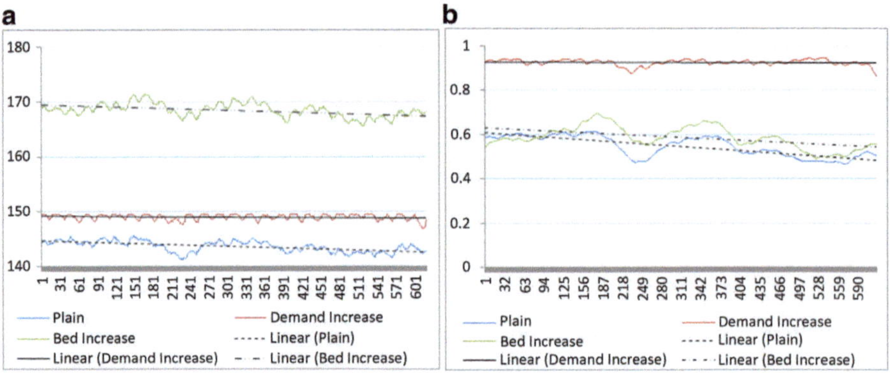

Fig. 9 Bed occupancy ratio (**a**) and elective blocking (**b**) in medicine wards

Conclusions

The present work develops a tool based on a system dynamics model aimed at understanding how the feedback mechanism among emergent and elective patient flows inside a hospital works in practice and ensures system equilibrium.

The simple process description is not enough to explain the capability of the system to face demand variations. Through the feedback loops introduced in the system dynamics representation it was possible to explain the mechanisms to balance erratic demand.

Moreover, the model evaluates the expected impact of alternative strategies directed to maintain the system's equilibrium if unpredictable peaks in ED activity occur (i.e., flu epidemic peaks).

An important contribution is also given by the assessment of the effects produced by the elective patients blocking, mainly caused by overcrowding.

In the case study presented, by increasing the inpatient bed capacity by approximately 20 %, an increase in ED arrivals of 60 % could be accommodated. In future works, we want to increase the boundary of simulated system, by including primary care facilities to better assess the flows of patients addressing the hospital and to study the system from a network perspective

Acknowledgements The authors acknowledge support from the Italian Ministry of Education, University and Research (MIUR), under the grant FIRB n. RBFR081KSB. Data were made available thanks to a collaboration between ARS Liguria (Dr. Francesco Quaglia and Domenico Gallo) and the Department of Economics and Business, University of Genova.

References

Abo-Hamad, W., Arisha, A.: Simulation-based framework to improve patient experience in an emergency department. Eur. J. Oper. Res. **224**(1), 154–166 (2013)

Aboueljinane, L., Sahin, E., Jemai, Z.: A review on simulation models applied to emergency medical service operations. Comput. Ind. Eng. **66**, 734–750 (2013)

Agenas (National Agency for Health Care Services): Monitor, Special Issue on the Emergency System. Rome (Italy). AGENAS, http://www.agenas.it/pubblicazioni/monitor-rivista-agenas (2012). Accessed 15 Feb 2015

Bagust, A., Place, M., Posnett, J.: Dynamics of bed use in accommodating emergency admissions: stochastic simulation model. Br. Med. J. **310**(7203), 155–158 (1999)

Brailsford, S.C., Lattimer, V.A., Turnbull, J.C.: Emergency and on-demand health care: modelling a large complex system. J. Oper. Res. Soc. **55**, 34–42 (2004)

Brailsford, S.C.: System dynamics: what in it for healthcare simulation modelers. In: Mason, S.J., Hill, R.R., Monch, L., Rose, O., Jefferson, T., Fowler, J.W. (eds.) Proceedings of the 2008 Winter Simulation Conference. Institute of Electrical and Electronics Engineers, New York (2008)

Brailsford, S., Churilov, L., Dangerfield, B.: Discrete-Event Simulation and System Dynamics for Management Decision Making. Wiley, Chichester (2014)

Duguay, C., Chetouane, F.: Modeling and improving emergency department systems using discrete event simulation. Simulation **83**(4), 311–320 (2007)

Günal, M.M., Pidd, M.: Discrete event simulation for performance modelling in health care: a review of the literature. J. Simulat. **4**, 42–51 (2010)

IHI: Optimizing Patient Flow: Moving Patients Smoothly through Acute Care Settings. IHI Innovation Series White Paper. Institute for Healthcare Improvement, Boston (2003)

Katsaliaki, K., Mustafee, N.: Applications of simulation within the healthcare context. J. Oper. Res. Soc. **62**(8), 1431–1451 (2010)

Lane, D.C., Monefeldt, C., Rosenhead, J.V.: Looking in the wrong place for healthcare improvements: a system dynamics study of an accident and emergency department. J. Oper. Res. Soc. **51**, 518–531 (2000)

Paul, J.A., Lin, L.: Models for improving patient throughput and waiting at hospital emergency departments. J. Emerg. Med. **43**(6), 1119–1126 (2012)

Proudlove, N.C., Black, S., Fletcher, A.: OR and the challenge to improve the NHS: modelling for insight and improvement in in-patient flows. J. Oper. Res. Soc. **58**, 145–158 (2007)

Simeu: Standard organizzativi delle strutture di emergenza e urgenza. Organizational standards of A&E facilities. http://www.comesemergenza.it/files/Standard_FIMEUC-SIMEU_2011.pdf (2011). Accessed 15 Feb 2015

Vanderby, S., Carter, M.W.: An evaluation of the applicability of system dynamics to patient flow modelling. J. Oper. Res. Soc. **61**, 1572–1581 (2010)

Wong, H.J., Morra, D., Wu, R.C., Caesar, M., Abrams, H.: Using system dynamics principles for conceptual modelling of publicly funded hospitals. J. Oper. Res. Soc. **63**(1), 79–88 (2012)

Markov Decision Process Model for Patient Admission Decision at an Emergency Department in Disasters

Hyun-Rok Lee and Taesik Lee

Introduction

In the aftermath of disasters, hospitals need to accommodate a large number of patients in a short period of time. Emergency Medical Service (EMS) systems undergo transient burdens to take care of the surge demand resulting. In such events, hospitals, emergency rooms in particular, experience shortage in medical resources needed for providing appropriate care. Efficient use of scarce resources is important in order to save more victims injured in a disaster.

There are many resource allocation decisions to be made during the response to a disaster. One example of such decisions is to determine the order of evacuating victims from a disaster scene to transport to a hospital. The most common practice is to allocate ambulances to victims in the order of severity in their medical condition. Simple Triage and Rapid Treatment (START) is one of the standard triage method to assess the severity of victims. By a few quick and simple tests, victims are classified into four categories: deceased/expectant, immediate, delayed, and minor (Jenkins et al. 2008). Then, the victims in immediate-class are assigned the highest priority for ambulance transport, delayed-class next, then the rest of the groups. While this is a standard practice in disaster medicine, a few recent studies point out this relatively simple approach may not be the optimal strategy (Argon et al. 2008; Jacobson et al. 2012; Jenkins et al. 2008; Mills et al. 2013; Sacco et al. 2005). These studies highlight the fact that higher medical urgency, represented by triage class, does not necessary imply a higher chance of survival when emergency care is provided; reduction in the chance of survival by delaying care for delayed-class patients may be larger than the gain from treating immediate-class patients.

H.-R. Lee • T. Lee (✉)
Industrial & Systems Engineering, KAIST, 291 Daehak-ro,
Yuseong-gu, Daejeon 305-701, Korea
e-mail: hyunrok@kaist.ac.kr; taesik.lee@kaist.edu

A. Matta et al. (eds.), *Health Care Systems Engineering for Scientists and Practitioners*, Springer Proceedings in Mathematics & Statistics 169,
DOI 10.1007/978-3-319-35132-2_18

This paper addresses a decision problem concerning the admission of a patient—whether to admit a patient into an Emergency Department (ED) and consume one of the remaining beds. Our research question has been motivated by the results of previous research which indicates that occupation of beds by low-severity patients may cause an increase in mortality owing to the resulting bed shortage for high-severity patients who arrive later (Auf der Heide 2006; Einav et al. 2006). High-severity patients do not always arrive before low-severity patients from the scene of a disaster. It is observed that they, in fact, usually arrive later, and selectively admitting patients can be an effective mitigation strategy for the temporal resource shortage problem (Auf der Heide 2006). In the most extreme form, a selective admission policy may state that low-severity patients (e.g., those in the minor category) should not be admitted to the ED in disaster situations. But such simple strategy does not have clear criteria of when to start and when to finish the protocol. More sophisticated decision making, guided by engineering modeling and analysis, can achieve better operation on patient admission control and save more lives in the event of disasters.

The problem of efficient use of limited resources is a much studied research topic in the field of revenue management, but little research has been conducted in the context of disaster response. Among a few examples of the prior research is Jacobson et al. (2012) where they solved a stochastic scheduling problem with plausible conditions on service time and life time in a disaster. Kilic et al. (2013) proposed a queuing model to characterize disaster situations and determined the optimal service level in response to surge demand experienced at a hospital. Cohen et al. (2014) constructed a continuous optimization problem to determine the optimal allocation of doctors between two treatment areas.

The objective of this paper is to develop an optimal policy for patient admissions to an ED for better utilization of scarce ED resources in the aftermath of disasters. We develop a Markov Decision Process (MDP) model to solve a multi-class admission control problem over a finite horizon with time-dependent arrival rate and time-dependent rewards. The MDP model developed in our research is in line with the model of Brumelle and Walczak (2003) which solves a revenue maximization problem for a single flight through admission control of multiple fare classes with time-dependent arrivals to a single-leg flight. Our model considers an additional constraint that patients in high-severity patient classes should not be rejected from an ED as long as there is a remaining capacity. We implement a discretization technique to get numerical solutions to the MDP model. Optimal policies obtained from this model can help EDs to efficiently utilize scarce resources (i.e., beds in EDs) to save more victims in the event of disasters.

MDP Model

Our problem is to determine the admission of patients arriving at an ED in the aftermath of a disaster. We assume based on the previous research (Mills et al. 2013) that the condition of a patient continues to deteriorate until definitive care (e.g., trauma surgery) is provided. In terms of MDP modeling, it implies that the patients have a time-dependent reward. We consider M patient classes, and assume that there is a particular patient class, *Priority*-class, that should be always admitted to the ED as long as there is a remaining bed in the ED. For example, in the START triage practice, there are four patient classes, and patients in the immediate-class may be given the priority for admission. For the arrival process, we assume that patients from each class arrive according to an independent Nonhomogeneous Poisson Process (NHPP). This is to represent time-varying arrivals of patients to an ED from a disaster and to incorporate different arrival patterns between patient-classes. In our model, we only consider beds in an ED as the target for resource allocation, and it is assumed that once a bed is given to a patient, it will not become available during the time horizon. In other words, we assume that the decision time horizon is short compared to treatment time for patients and no admitted patients leave the ED within that time horizon. We solve this problem by using an MDP model.

An MDP is a sequential decision-making methodology that finds the optimal action for each system state (Puterman 2009). In our model, the state of the system is defined as a vector (k, t_i, m), where k is the number of remaining beds in the ED, t_i is the time elapsed from the onset of a disaster, and m is the class of the currently arriving patient ($m \in M$). We discretize the entire time horizon into N time interval with uniform length, Δt. The admission decision in state (k, t_i, m) concerns an arriving patient during time interval $(t_{i-1}, t_i]$. It is assumed here that only one patient can arrive in a time interval. As a consequence, a decision epoch in the MDP model is the nearest decision time after the moment of patient arrival. In other words, an admission decision for an arriving patient is made at a predefined point, rather than the moment of the patient's arrival, by delaying the decision till the end of the time interval. Actions in our model represent the admission decision for the currently arriving patient—$a = \{\text{admit patient, reject patient}\}$. Recall that we assume that patients from *Priority*-class are always admitted. That is, if there is a remaining bed and an arriving patient is in *Priority*-class, only the "admit patient" action can be chosen.

Transition of states depends on the chosen action at the current state. The number of remaining beds k stays unchanged if the action for the currently arriving patient is "reject". Otherwise, the number of remaining beds is reduced by one, except in cases where the remaining bed was zero. We assume that beds in the ED can only be consumed and not restored—i.e., there is no patient leaving the ED once admitted. Note that in our model, in addition to the admission decision, state transition also depends on the patient arrival process.

We assume that patient arrivals from each class follow an independent NHPP. Let $\lambda_m(t)$ denote an intensity function for the NHPP for class-m patients, which indicates the arrival rate of class-m patients at time t. We also let $\lambda(t)$ denote the sum of the arrival rates of all the patient classes. By the property of NHPP, the probability that a patient who has arrived at t_j is from class-m is $\lambda_m(t_j)/\lambda(t_j)$, and the probability of a patient arrival during $(t_{j-1}, t_j]$ given an arrival at t_i is

$\{exp(-\int_{t_i}^{t_{j-1}} \lambda(s)ds) - exp(-\int_{t_i}^{t_j} \lambda(s)ds)\}$. Then, at current time t_i, the probability that a class-m patient arrives during $(t_{j-1}, t_j]$ is:

$$P_{t_j,m|t_i} \approx \frac{\lambda_m(t_j)}{\lambda(t_j)} \left\{ exp\left(-\int_{t_i}^{t_{j-1}} \lambda(s)ds\right) - exp\left(-\int_{t_i}^{t_j} \lambda(s)ds\right) \right\}. \quad (1)$$

Hence, the transition probability from state (k, t_i, m^o) to the next state, $([k-1]^+, t_j, m)$, for example, is $P_{t_j,m|t_i}$ if the chosen action for the currently arriving patient is "admit" or 0 if "reject" is chosen.

The value of state (k, t_i, m) is the maximum expected sum of the reward and future state values. We define $r_m(t_i, a)$ to be the reward received when a class-m patient arrives at the ED at time t_i. Non-zero reward is gained even if we reject a patient, but the reward amount is discounted compared with the reward when the patient is admitted to the ED. More discussion on the reward function is presented in section "Model Inputs". The value function is set to zero if the elapsed time of the next state is greater than t_N, which limits the planning horizon. Another boundary condition is that the value of the states whose $k = 0$ (i.e., no bed) is zero because we cannot admit any patient in those states and their future states. The value function is recursive and has the form of the Bellman equation as follows:

$$V(k, t_i, m) = \max_a \left[r_m(t_i, a) + \sum_{\sigma \in M} \sum_{j=i+1}^{N} \{P_{t_j,\sigma|t_i} \times V([k - 1_{admit}]^+, t_j, \sigma)\} \right]. \quad (2)$$

where 1_{admit} is an indicator function, and it is 1 if a is "admit patient" and 0 if a is "reject patient", and $i = 1$ to N.

Since the above, discretized MDP model has a finite state space, it can be easily solved using a backward iteration algorithm. This algorithm recursively calculates the value of each state using pre-calculated values of possible next states. State transition of our MDP model only proceeds to smaller k and larger t_i. Therefore, the value of a state can be calculated from the states closer to the end of the planning horizon t_N that have a smaller k. We calculate the optimal value from state $(1, t_N, m)$ for $m \in M$, to state $(1, 0, m)$ for $m \in M$, then state $(2, t_N, m)$ for $m \in M$, and so on. The boundary condition is that the value of the states with $k = 0$ or $t_i > t_N$ is zero.

Computational Result

We compute a solution of the MDP model constructed in the previous section and test its performance. In this section, we first explain the input functions used in our MDP model, which are adopted from the previous literature. Then, we briefly discuss a general structure of the policy obtained from the MDP model, followed by experimental results to compare the policy's performance with a simple, first-come-first-serve policy.

Model Inputs

In our test, we consider four patient classes from the START triage—immediate, delayed, minor and expectant. Immediate-patients require definitive care be provided as quickly as possible for their survival. Delayed-patients also require emergency care as soon as possible, but they can wait a longer period before their condition gets critical, relative to immediate-patients. Minor-patients are without life-threatening injuries (e.g., minor fracture), and care for these patients is not time-critical. Expectants are either found or expected to be deceased at a disaster scene. To simplify the analysis, we specifically focus on the immediate- and delayed-patients. This simplification is in line with a standard practice where we do not use emergency medical resources for minor- or expectant-patients until all patients in the immediate and delayed class are treated. We assume that immediate-patients are never rejected for their admission to an ED, which is also in line with a common ED practice. With this assumption, our problem becomes to determine an admission decision for delayed-patients given the current state.

We use NHPP to characterize the arrivals of patients to an ED. Following Kilic et al. (2013), the intensity functions for the NHPP are assumed to have a modified form of the probability density function of gamma distribution as follows:

$$\text{Arrival rate} = \frac{C}{60}\frac{(t/60)^{\alpha-1}\exp(-t/60\beta)}{\Gamma(\alpha)\beta^{\alpha}} \tag{3}$$

Parameters α and β have been chosen such that they demonstrate the patient arrival patterns known to exist in disasters; (1) patient arrivals rapidly increase at the beginning, reaches its peak, and decreases to zero, and (2) delayed-patients show up at an ED earlier than immediate-patients. Specifically, we choose $\alpha = 2.5, \beta = 1.0$ for λ_I and $\alpha = 1.5, \beta = 1.0$ for λ_D. In our experiments, we test three cases of disaster scales by changing a scaling constant in the arrival rate functions. The expected volume of patients from a disaster (during the 12 h after the disaster) are set at the level of 30, 60, and 90 for each case with the ratio between immediate- and delayed-patients is 1–3 in all three cases.

For the reward function for the two classes of patients, $r_I(t,a)$ and $r_D(t,a)$, we adopt the survival probability functions, $f_I(t)$ and $f_D(t)$, proposed in Mills et al. (2013). The critical feature in the survival probability functions is that the rate of decrease is higher for $f_I(t)$ upto certain t and then the rate for $f_D(t)$ becomes higher after that point. To examine the effect of time-dependent reward on the optimal policy, we also test a time-independent reward function where we assume the reward for treating an immediate-patient is four times greater than treating a delayed-patient: $f_I = 4 \cdot f_D$.

The survival probability functions $f_I(t)$ and $f_D(t)$ define the reward when a patient is admitted. When patients are rejected, we use a discounted amount of reward. Note that a patient in the immediate-class can also be rejected when the ED is fully occupied. When rejected, we assume that those patients are diverted to other EDs in the region to receive necessary care. To account for the delay in care provision due to diversion, we discount the reward values from the original $f_I(t)$ and $f_D(t)$. We use two discount scheme, which we call as a degradation penalty and time-shift penalty. For a degradation penalty, we discount the survival probability by multiplying a discount factor—0.1795 for $f_I(t)$ and 0.0404 for $f_D(t)$ respectively. These factors are assumed based on the mortality rate for the interhospital transfer patients (Korea National Emergency Medical Center 2007). For a time-shift penalty, we directly account for the (expected) delay in care provision by adding 50 min for interhospital transfer time (Korea National Emergency Medical Center 2007). That is, when a patient is rejected from the ED at time t, the reward function for the reject action is given by $f_I(t+50)$ for immediate-patient and $f_D(t+50)$ for delayed-patient.

Policy Structure

We have three arrival scenarios—30, 60, and 90 patients—for which three types of reward functions are used—time-independent,[1] time-dependent with degradation penalty, and time-dependent with time-shift penalty. An optimal policy is obtained for each of the nine test instances.

Across all test instances, the optimal policies exhibit a common structure. There exists a distinct decision boundary between admitting and rejecting a delayed-patient, as illustrated in Fig. 1. The region above each curve represents the state space in which admitting a delayed-patient is an optimal action. When the system is in states below the curves, it is optimal to reject a delayed-patient.

Most of the optimal policies from the test instances exhibit a pattern shown in Fig. 1a; a switch in optimal action—from rejecting to admitting a delayed-patient—occurs earlier in time when there are a higher number of remaining beds in an ED. For example, for the curve for 90-arrivals, the switch occurs at around 200 min when the number of remaining beds is 4, whereas it occurs at around 100 min when there are 8 beds remaining in the ED. This has an intuitive explanation. Recall that the reason to reject a delayed-patient is to preserve a bed for an immediate-patient that may arrive in future. Also recall that arrivals of immediate-patients are time-dependent such that fewer patients arrive as longer time elapses. Thus, when it is 200 min after the disaster compared with 100 min after the disaster, fewer immediate-patients are expected to arrive in future, hence fewer number of beds needs to be saved for them. Note that reserving beds for immediate-patients makes sense only when the immediate-patients are expected to arrive *soon* enough; otherwise, the reward from using the beds for those future immediate-patients will become too small to justify rejection of the delayed-patients. If we compare the three curves of 30-, 60-, and 90-arrivals, we see that at the same elapsed time (e.g., approximately 100 min), the number of remaining beds for the switching increases

[1]For a time-independent reward function, we assume rejecting a patient results in zero reward.

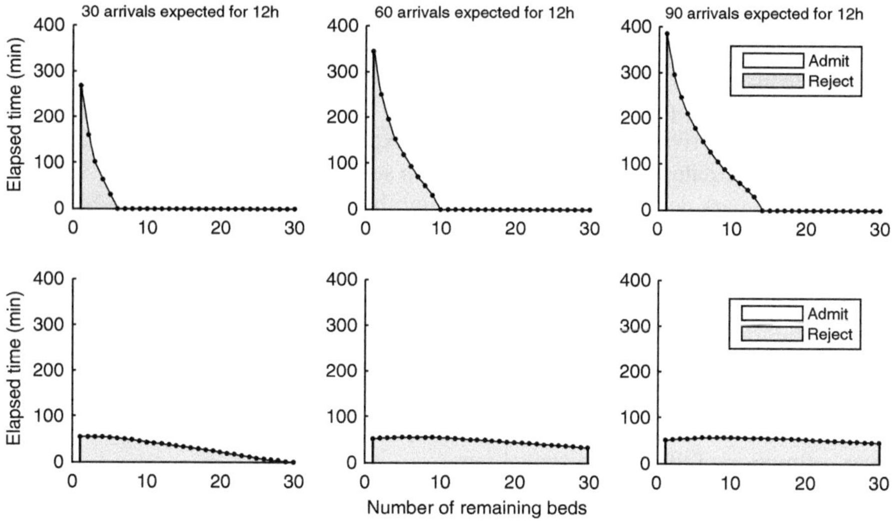

Fig. 1 Structure of a policy. (**a**) Time-dependent reward with degradation penalty. (**b**) Time-dependent reward with time-shift penalty

from 3 for 30-arrivals, 6 for 60-arrivals, to 8 for 90-arrivals. For a small-scale disaster (i.e., 30 patients in total), when an ED has 3 beds remaining at the moment, the ED does not need to reject a delayed-patient because it has sufficient beds to accommodate future immediate-patients. For a large-scale disaster of 90-arrivals, a sufficient capacity level changes to 8 beds.

There are a few instances where the optimal policy exhibits a pattern that is not quite intuitive. Figure 1b shows optimal policy curves for such cases where we have a time-dependent reward function with time-shift penalty. While in Fig. 1a, the boundaries monotonically decreases, they have a peak in Fig. 1b. Examining the curve for 90-arrivals at approximately 55 min of elapsed time, we see that an optimal action switches from "admit" delayed-patients to "reject" (3 beds) and back to "admit" (16 beds). This may be due to the fact that reward functions with time-shift penalty are not monotonically decreasing functions unlike in other cases.

Performance Test

Performance of the optimal policy from the MDP model is compared to other decision strategies by using virtual patient arrival data. For the comparison, we use an optimal MDP policy computed for the 60-arrival case as our reference MDP policy. Two decision strategies are considered as benchmarks: *first-come-first-serve* (FCFS) and *reject-delayed* (RD) strategy. In the FCFS scheme, an arriving patient is admitted regardless of her class, as long as there is a remaining bed. RD strategy rejects all delayed-patients and only admits immediate-patients until all beds are used up. This strategy is the simplest form of selective admission strategy in disaster management.

To create virtual patient arrival data, we use the acceptance-rejection method to generate NHPP arrival data of 30, 60, and 90 expected arrivals during the 12 h. One hundred sample datasets are analysed for each setting to compare the MDP solution to two benchmark decision strategies. The number of beds available in an ED at the beginning of the disaster is set at 30, 15, and 5 to create three levels of resource scarcity. Performance of these three decision strategies is measured by the total reward earned during the horizon. It is measured in terms of a relative efficiency against the ideal solution, where the ideal solution is the admission decision when patient arrival information—arrival time and patient type—is known a priori. Note that the MDP policy has been obtained for the 60-arrival case, and thus for 30- and 90-arrival samples, it is not the optimal policy but suboptimal due to under- or over-estimation of the patient arrival rate.

Table 1 shows the results from the 60 expected arrival samples. It shows the relative efficiency for the three decision strategies under three reward functions. The three reward functions are (1) time-independent reward (TI), (2) time-dependent reward with degradation penalty (TDDP), and (3) time-dependent reward with time-shift penalty (TDTS). The results show that the MDP policy is better than the two benchmarks in all settings. Sample mean of the MDP policy results is significantly different from FCFS (9 sets in 9 sets) and RD (6 sets in 9 sets) at 5 % significance level. FCFS strategy performs better when more beds are initially available at an ED. This is because more beds imply less incentive to reject delayed-patients and thus FCFS works relatively better compared with when there are fewer beds to begin with. RD (reject-delayed) strategy, on the other hand, shows the opposite trend: its performance improves when fewer beds are available. Since RD strategy rejects all delayed-patients, it sometimes ends up leaving some beds unused while delayed-patients are rejected. When only five beds are available, they tend to be filled up quickly by immediate-patients, hence RD strategy suits well. The MDP policy performs remarkably well in all settings with a minimum of 96.93 % and maximum of 100 % efficiency.

Computing an optimal policy from the MDP model requires knowledge or assumption on the patient arrival rate, whereas the FCFS and RD strategy do not require such assumption. It is unrealistic to expect accurate information on the arrival rate is available in disasters, and thus it is important for the MDP policy to be reasonably robust against the uncertainty in the patient arrival rate. Table 2 shows the results from the tests where the expected patient arrival rate of the sample data is 30 and 90 as well as 60. It shows that the relative efficiency of the MDP policy is fairly high even when the expected arrival rate assumed in the MDP model is different from the sample data. The relative efficiency is over 90 % for most cases with the minimum value of 87.59 %. This result demonstrates the possibility that optimal policies from our MDP model can be reasonably applied in the presence of uncertainty in predicting the patient arrival rate in the event of disasters.

Table 1 Relative efficiency of three different policies

Reward function	Strategy	Initial # of beds 30	15	5
TI	FCFS	57.95 (1.79)	34.14 (1.35)	27.25 (1.35)
	RD	77.59 (1.56)	96.38 (0.95)	100 (0)
	MDP policy	96.93 (0.60)	97.79 (0.59)	100 (0)
TDDP	FCFS	94.19 (0.67)	79.65 (1.36)	54.92 (1.86)
	RD	53.98 (2.21)	85.93 (2.34)	99.86 (0.26)
	MDP policy	99.04 (0.20)	98.17 (0.39)	99.86 (0.26)
TDTS	FCFS	85.46 (1.17)	66.35 (1.59)	48.67 (1.71)
	RD	23.13 (1.36)	42.43 (2.29)	75.21 (2.80)
	MDP policy	98.12 (0.29)	97.78 (0.31)	97.27 (0.49)

The numbers in the parentheses are 95 % confidence interval

Table 2 Relative efficiency of the MDP policy to the variation of the expected arrival volume in the data

Initial # of beds	30			15			5		
Expected volume	Less	Same	More	Less	Same	More	Less	Same	More
TI	90.60	96.93	94.96	89.11	97.79	99.80	99.17	100	100
	(0.91)	(0.60)	(1.03)	(1.58)	(0.59)	(0.18)	(0.72)	(0)	(0)
TDDP	99.40	99.04	97.88	94.29	98.17	96.4	94.77	99.86	100
	(0.20)	(0.20)	(0.39)	(0.96)	(0.39)	(0.56)	(1.70)	(0.26)	(0)
TDTS	87.58	98.12	97.10	92.34	97.78	98.04	96.53	97.27	97.89
	(1.23)	(0.29)	(0.33)	(1.75)	(0.31)	(0.60)	(0.81)	(0.49)	(0.40)

The numbers in the parentheses are 95 % confidence interval

Conclusion

Efficient use of scarce resources is a critical decision problem in disaster management. In this paper, we study one of such decision problem, which is to make an admission decision for patients arriving at an emergency department. Specifically, we consider a case where patients from two classes of severity—immediate and delayed—arrive by independent NHPPs and their survival probabilities decrease as a function of time. A finite horizon MDP model is constructed to derive an optimal policy, which is obtained by solving the discretized model by using a backward iteration algorithm. An optimal action is defined for each state (k, t_i, m) where k is the number of remaining beds in the ED, t_i is the time elapsed from the occurrence of a disaster, and m is the triage class of the currently arriving patient. Experimental results show that optimal policies obtained from the MDP model outperforms the two benchmark decision making strategies which are in practical use in EDs. The optimal policies performs reasonably well even when an assumption for patient arrivals is not exactly correct. This is important because in disaster scenarios, accurate information on patient arrivals is not available. The optimal policies derived in this research needs further investigation for their structural property.

Acknowledgements This research was supported by a grant 'research and development of modeling and simulating the rescues, the transfer, and the treatment of disaster victims' [nema-md-2013-36] from the man-made disaster prevention research center, national emergency management agency of korea.

References

Argon, N.T., Ziya, S., Righter, R.: Scheduling impatient jobs in a clearing system with insights on patient triage in mass casualty incidents. Probab. Eng. Inf. Sci. **22**(3), 301–332 (2008)

Auf der Heide, E.: The importance of evidence-based disaster planning. Ann. Emerg. Med. **47**(1), 34–49 (2006)

Brumelle, S., Walczak, D.: Dynamic airline revenue management with multiple semi-Markov demand. Oper. Res. **51**(1), 137–148 (2003)

Cohen, I., Mandelbaum, A., Zychlinski, N.: Minimizing mortality in a mass casualty event: fluid networks in support of modeling and staffing. IIE Trans. **46**(7), 728–741 (2014)

Einav, S., Aharonson-Daniel, L., Weissman, C., Freund, H.R., Peleg, K., Israel Trauma Group.: In-hospital resource utilization during multiple casualty incidents. Ann. Surg. **243**(4), 533–540 (2006)

Jacobson, E.U., Argon, N.T., Ziya, S.: Priority assignment in emergency response. Oper. Res. **60**(4), 813–832 (2012)

Jenkins, J.L., McCarthy, M.L., Sauer, L.M., Green, G.B., Stuart, S., Thomas, T.L., Hsu, E.B.: Mass-casualty triage: time for an evidence-based approach. Prehosp. Disaster Med. **23**(1), 3–8 (2008)

Kilic, A., Dincer, M.C., Gokce, M.A.: Determining optimal treatment rate after a disaster. J. Oper. Res. Soc. **65**(7), 1053–1067 (2013)

Korea National Emergency Medical Center, Ewha Womans University: Development of the Guidelines for Inter-Hospital Transfer. Korea National Emergency Medical Center, Seoul (2007)

Mills, A.F., Argon, N.T., Ziya, S.: Resource-based patient prioritization in mass-casualty incidents. Manuf. Serv. Oper. Manag. **15**(3), 361–377 (2013)

Puterman, M.L.: Markov Decision Processes: Discrete Stochastic Dynamic Programming. Wiley, New York (2009)

Sacco, W.J., Navin, D.M., Fiedler, K.E., Waddell, I.I., Robert, K., Long, W.B., Buckman, R.F.: Precise formulation and evidence–based application of resource–constrained triage. Acad. Emerg. Med. **12**(8), 759–770 (2005)

Crisis Management Plan: Preventive Measures and Lessons Learned from a Major Computer System Failure

Hélène Grange and Jérémie Leynon

Introduction

Computers are omnipresent within hospitals. All possible measures are taken to avoid failure. It is the responsibility of the IT department to ensure the security of data. On August 28, 2014, Saint-Joseph/Saint-Luc Hospital was confronted with a major computer system failure. The hospital is a private non-profit establishment with a capacity of 340 beds for in- and outpatients. It has a variety of specialist medical and surgical departments, including a burns unit, an intensive care unit, an emergency department and a maternity unit. It records approximately 22,000 hospitalizations per year, 90,000 appointments and 35,000 visits to the emergency department.

The hospital activities are organized per process. In Fig. 1, we find the various medical specialties required by patients. We can also see various medical departments called upon to meet these special requirements, such as medical biology, blood banks, conventional and interventional imagery, cardiological exploration, and operating theatres. There are also various support services, such as the pharmacy, patient administration, social support, internal transport, infection risk management, and patient records. Other support services include reception, accommodation, logistics, information systems, maintenance of facilities and biomedical equipment, waste management, procurement, and hospital safety and security. The strategic management of the hospital and coordination of activities is carried out by the general management department and the hospital departments. An executive

H. Grange (✉) • J. Leynon
Centre Hospitalier Saint-Joseph/Saint-Luc, 20 Quai Claude Bernard,
Lyon Cedex 07 69365, France
e-mail: hgrange@ch-stjoseph-stluc-lyon.fr

A. Matta et al. (eds.), *Health Care Systems Engineering for Scientists and Practitioners*, Springer Proceedings in Mathematics & Statistics 169,
DOI 10.1007/978-3-319-35132-2_19

board composed equally of doctors and hospital department managers defines this strategy. A committee composed of hospital managers from all departments implements the guidelines.

The Event

Onset of the Computer System Failure

As the department manager on call on August 28, I received a call at 10.30 p.m. from a biology lab technician indicating that she could no longer send test results to patients' medical records or receive requests for examinations. I therefore called a manager in the information systems department, who attempted in the first instance to diagnose and resolve the problem remotely, without success.

Given the situation, we decided to meet at the hospital to conduct a thorough diagnosis and attempt to reboot the computer rack in question. At midnight, we obtained a diagnosis: a number of software programs had shut down, including the administrative patient management software; the identity, movements and invoicing servers; the software used for medical records, nursing records, and the drug circuit (prescription and drug administration); specific software for the management of surgical theatre equipment required to be sterilized for operations; the stock purchasing program; contracts, general and cost accounting; the biology laboratory information system; the chemotherapy preparation software system; all imaging software; the payroll software (see Fig. 1).

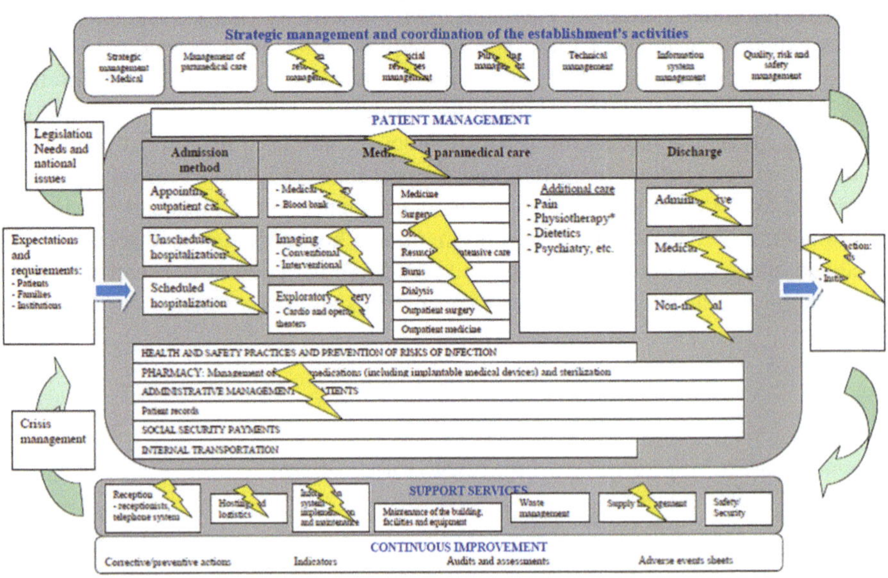

Fig. 1 Map of the hospital's processes and impacts of computer system

In Failure Mode

It was impossible to admit and register any new patients to the emergency department, or to transfer files containing the medical information required by inpatient services. It was therefore decided, with the emergency physicians, to put a stop to all admissions to the emergency unit by contacting the Lyon medical emergency transport services.

At 1.00 a.m., I decided to start the "paper kit" procedure and called in the health care manager. Each department or entity that uses business software applications must continue to function in the absence of these applications (degraded mode). The failure mode procedure had been set up by a group from the hospital working on emergency plans. The group worked on a "paper kit" for all departments to ensure the continuity and traceability of care. To launch this system, the IT department had set up an automated system to record the latest treatments and drug administration (carried out every 4 h). Operation in degraded mode implies: a smooth transition from nominal operation mode to the degraded mode; a proper operation in degraded mode for the duration of the system failure; a smooth return to nominal operation once the system becomes again available.

We decided to print all hospital documents in the "paper kit" and distribute them to all departments to ensure continuity of care.

At that time of the night, we had no idea when a return to nominal operation could be expected, and we had to explain to the medical staff how and what to do.

At 4 a.m., our IT department contacted the IT support company located in the United States in another attempt to restore the system without losing the medical data contained on the two failed disks.

At 6 a.m., we completed our visits to the various departments and I set up a crisis meeting for 8 a.m. The president of the governing medical committee (who is also the physician in charge of the emergency department), the healthcare manager, the information systems director, the head of anesthesia, the head of the biology laboratory, the manager of the bed reservation centre, the appointments manager, the head of administrative patient management, the CFO, the managers of the medical departments, and the logistics manager, were present. We considered the following points: repair of the hospital information system; administrative management of patients; access to patient prescriptions and medication; administration of records; impact on the biology laboratory, operating theatres, appointments, imaging and the blood bank.

At 9 a.m., the initial crisis report was completed and sent by email to all staff and displayed at all nurses' stations. A backup calendar of appointments was printed along with other paper documents, such as observation sheets, but physicians had to carry out appointments without access to patients' medical records. We asked the secretarial department to make a note of the contact details of patients calling to make appointments, so that we could schedule appointments once the hospital information system was operational again.

In terms of hospitalizations, when staff arrived at 7 a.m., they were informed of the failure and the launch of the "paper kit" procedure for prescriptions, patient monitoring, traceability of drug administration, and test requests. The pharmacy had to review its entire organization in order to deal with the disruption caused by the computer failure. We asked the departments to limit laboratory tests, and nurses had to go to the laboratory to retrieve test results. Radiology exams had to be requested by fax and a system of paper records implemented in the imaging department. The failure of the chemotherapy software meant that patients' treatments could not be prepared and a number of treatments had to be postponed.

We proposed to hold another crisis meeting at midday. At 2.00 p.m., we alerted the regional health agency. At 3.00 p.m., the IT department informed us that the repair process had failed; BULL would have to provide us with new discs from their Paris branch on Saturday morning. We would start afresh with a new storage array and we have to restore all of our backups. At 5.00 p.m., another crisis meeting was organized: we decided to cancel certain operations scheduled for the operating theater on Monday morning and maintain the suspension of surgical emergencies.

In the late afternoon, we decided to call all of the doctors on duty on Saturday and Sunday, and to hold another crisis meeting the following day. At 7.00 p.m., BULL confirmed that we would have to reinstall every application as if for an initial set-up. We requested their assistance with the restoration operation.

On Saturday, the storage array was finally repaired and reformatted. In the meantime, we encountered a high level of risk in terms of identity vigilance due to handwritten patient notes.

On Sunday, another crisis meeting was organized. The biology lab was still working in degraded mode and we asked that only urgent tests be prescribed. For patients already created in the electronic health records, new observations, correspondence and reports could be re-entered in the computer. Requests for examinations from the various technical platforms still had to be made on paper. We maintained the suspension of surgical emergencies. For new patients, the degraded operating procedure on paper was still ongoing.

On Monday, September 1, we maintained degraded mode for new patients. We set up a system to collect test results and have them brought to the care units. At the end of day, the application for making patient appointments was once again operational. Drug prescriptions still had to be carried out according to the degraded procedure. Chemotherapy activities were limited and patients' treatments were postponed to a later date. An appointment was made with our insurer. At 8.00 p.m., all applications were reinstalled but were not yet operational.

On Tuesday, restoration of the biology laboratory data was started. Chemotherapy activities ran as normal, but surgical emergencies were still suspended. Updates and diagnoses for patients already in the system could be entered in their computer records. In the afternoon, the import of the biology laboratory data was completed.

On Wednesday, the integrity of the laboratory database was checked and in the early afternoon, the application was operating correctly. The software interface between the biology laboratory and the departments was not yet operational; the degraded mode of operation was still being used. Other applications were also being restored and patient registration in the administrative software was once again possible. However, we could not make use of it, as we had to respect the chronology of entries from the beginning of the failure. The process to enter historical data would take several days. We called in reinforcement administrative staff in order to optimize the process and be able to resume current admissions with as little delay as possible.

On Thursday, September 4, we were finally able to resume surgical emergencies.

On Friday, September 5, all applications were operational and data was restored.

Resuming the use of electronic health records would take place in two stages: firstly, patients from Monday, 8 September, and subsequently, all patients from the following days. We asked that all paper documents be archived in case of medico-legal problems.

The week of September 8 was devoted to the recovery of non-entered data and the Medical Information Department did a lot of work on the detection of duplicate files. By Tuesday, September 9, the hospital was operating normally.

On Wednesday, we worked on an evaluation of the costs of the failure.

Throughout the failure, staff was concerned about the consequences of the failure on patients. Their primary question was whether they could take care of patients without risks and how to ensure a satisfactory level of service. The business continuity plan was organized quickly as the event occurred during the night; the crisis meeting was organized 10 h after the event. The availability of staff with the necessary computing skills for the recovery system must be improved. Staff adapted to using a paper-based system, but we encountered great difficulties in terms of patient identity. Medication management was particularly complex as our prescription software is interfaced with drug inventory management software; the supply of medications had to be based on a visual review of stocks. Reading prescriptions and dispensing drugs proved difficult. The biology laboratory was heavily affected and we had to ask the hospital departments to limit prescriptions. The failure generated a significant additional workload. Not all test results were collected by the prescribing departments and having to ensure their collection or distribution added greatly to the laboratory technicians' workload.

Governance

A crisis meeting has been organized daily from August 28 until September 4. It is composed of the main decision-makers: the general director, the executive assistant, the president and the vice-president of the governing medical committee, the health-care manager, the procurement director, the responsible of administrative patient management, the CFO, the physician in charge of biology laboratory, the physician in charge of the imaging service, the doctor responsible for the surgery, the doctor in charge of the anesthesia, the doctor responsible for medicine, the responsible for consultations. It decides on critical assets management: the restoration of the information system, the administrative patient management, the patient records, the prescriptions, the biology laboratory, the planning of operating rooms, the pharmacy, the chemotherapy, the consultations, the imaging, the communication, and their interlinking. Administrative supervisors, medical responsible, nursing responsible, and all professionals apply the degraded procedures and the instructions which are transmitted by the crisis unit.

Impact and Consequences

There was a high level of mobilization among staff and all of the department heads assumed the necessary responsibilities. Everyone had to reconsider their way of working, while continuing to accomplish their daily tasks. Over 80 % of staff was affected, including doctors, nurses, medical and technical professionals, logistics agents, administrative and medical secretariats. Staff was running out of steam on the sixth day of the failure and the management had to motivate and encourage them in their new practices. A major psychological impact was observed in staffs, who were disconcerted by this change from their usual practices.

The impact on appointments was limited, except on the first day, when doctors had to conduct appointments without patient records. The biology laboratory was severely disrupted and the return to normal activities was long and tedious. Certain test results were not requested, there were identity-vigilance problems, and technicians had to work directly on machines. In the care units, staff adapted well to a paper-based system but had difficulties in reading drug prescriptions. The radiology department was only slightly affected. Chemotherapeutic activities were severely affected; only mono-therapies could be prepared and patients requiring multi-therapy had to be rescheduled. The emergency department was affected, in that we had to suspend surgical emergency admissions. Also, the pharmacy was severely affected in that the preparation of required drugs was complicated by the poor legibility of prescriptions and last-minute adjustments by physicians. The paper system was not an appropriate solution and, consequently, there were long delays in obtaining the drugs prescribed. This part of the paper system must therefore be revised. Lastly, it looks a long time to recover patients' administrative data, because each additional day of the failure resulted in more patients being added to the list to be updated on the computer.

Lessons for the Future

The complete computer system failure procedure was not finalized and was known only to members of the Crisis Management Working Group. Paper documents were ready but staff had not been trained in their use. This system nevertheless allowed traceability of care and requests for tests to be maintained without too much difficulty. Electronic patient records were partially restored from a backup in the first 24 h, making it possible to find the most recent treatments, observations, nurses' transmissions and staff schedules.

The first threat is the failure to take into account the extent to which computer systems have penetrated our processes. The system failure resulted in a loss of confidence among staff, indicating the importance of the reliability of the information system. The second threat is the loss of integrity of data, which, fortunately was not the case at St Joseph/St Luc Hospital. The third threat is the loss of data availability; the total or partial lack of health-related data forced us to revert to basic methods of obtaining data. No exercises simulating a computer failure had previously been carried out, though we had planned for a carc unit to test the degraded operating process. Support for and communication with staff was essential not only during the crisis, but also afterwards. Certain staff required psychological support following the failure. Security must be considered beforehand, not after the event, and it is difficult to reassure staff that we are taking it into account when the system works correctly. Even residual risks can lead to a loss of critical resources that could cause disruption to activities.

Contribution of the ACP and ARP

The ACP is the Activity Continuity Plan. The main objective of the ACP is to ensure the availability of information regardless of the problems encountered. Unlike the ACP, the ARP (Activity Recovery Plan) does not ensure total availability of information. The ARP only ensures that activities may be resumed within in a predefined time.

Although organizations vary in terms of size and activities, they are all subject to scenarios that endanger their activities in the event of a threat to their computer systems. The hospital crisis recovery plan is therefore presented as a backup plan and a definitive way to restore normal operations. In the event of an accident or incident, the crisis recovery plan significantly reduces the financial consequences. It is necessary in order to safeguard an organization's activities, and also allows the organization to strengthen its image in terms of reliability and credibility with regard to its customers and partners. In general, the purpose of an ARP is to provide each organization with the best strategy to enable it to sustain its activities in the event of an incident, crisis or failure, and to subsequently resume operations within a reasonable time. Preventive measures are methods designed to avoid any kind of interruption. After determining the possible types of risk and their impacts, preventive measures can, for example, ensure that data is safeguarded through the use of solutions, backup systems, backup sites, which may be remote or on-site.

Remedial measures relate to the resumption of an organization's activities after a crisis. Many types of measures can be initiated, depending on the type of incident and its criticality. While a data backup system ensures the restoration or recovery of data, it is also possible to use a backup site.

Information System Risk Analysis

Following this traumatic incident and to meet the requirements of the French national "digital hospital" program, the hospital has initiated a comprehensive approach to management risks related to the information system. This approach comprises three aspects:

- Identify the most critical risks and implement the associated preventive and corrective actions.
- Define and deploy an information system security policy based on risk management.
- Create, formalize and test a plan for the continuity and resumption of activities in the event of incidents involving the information system.

Achieving and maintaining these three objectives would provide a partial solution to such traumatic incidents.

Risk Management Approach

This approach was initiated in September 2013, based on a group of 21 Rhône-Alpes institutions, accompanied by a consulting firm (Cosialis-Soluzen Consulting) and supported financially by the ARSRA (Agence Régionale de la Santé Rhône-Alpes). But, the question remained of how to perform a risk analysis without having to start from zero. For this, repositories are available: ISO 27005, AFNOR 2015 (Information security risk management), the repository of threats to information systems EBIOS, ANSSI 2015 (expression of needs and identification of security objectives), ISO 27002, AFNOR 2015 (Code of practice for information security management).

Figure 2 presents the risk management framework for the approach proposed by the consulting firm. It is based on the following steps:

- Define evaluation criteria: Quality of care, financial consequences, institution's image, the legal consequences, and the organization, are the selected criteria.
- Define criticality scales: the scales of value for each criterion, the scale of likelihood, the risk calculation formula, the action threshold for a risk, are specified.
- Create an incident scenario: such as a computer-room conflagration, a cyber attack…
- Calculate the risk: list the critical assets involved by the risk, define impacts and consequences on critical assets, calculate the likelihood, define the countermeasures, and calculate the residual risk with countermeasures.
- Establish security needs: regroup countermeasures on critical assets.
- Establish risk management: plan the countermeasures to implement on critical assets.

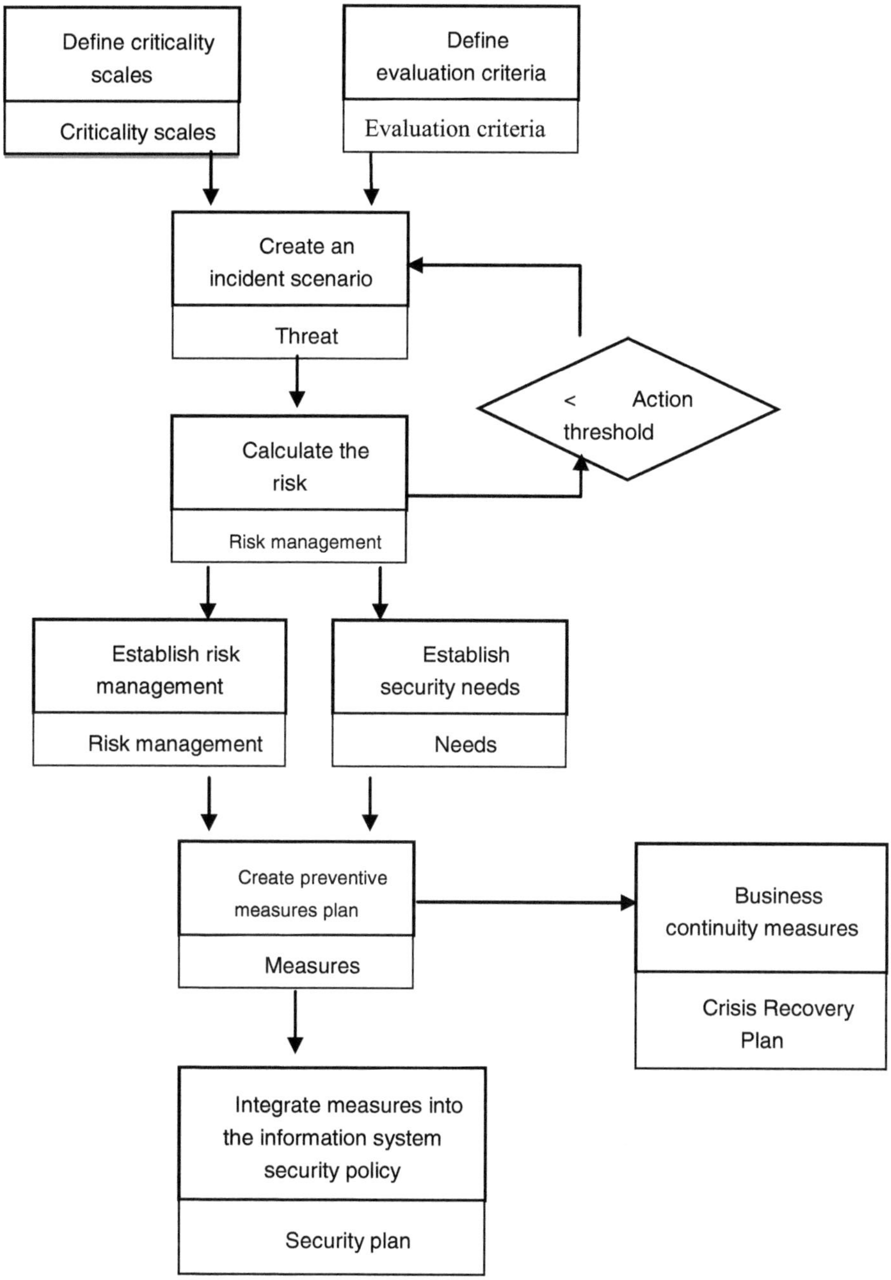

Fig. 2 Risk management framework

- Create preventive measures plan: define a risk management plan integrating the countermeasures.
- Business continuity measures: define a recovery plan for residual risks.
- Integrates measures into information system policy: integrate the countermeasures and the plans, in the information system policy.

The priorities set out by the framework, are as follows:

- Map the information system. This entails to reference all the elements comprising the information system, the application servers and the links between them.
- Identify activity processes and identify the connections with electronic applications.
- For each activity process, define the AICT (availability, integrity, confidentiality and traceability) requirements.
- Measure the impact of requirements on: legal and financial constraints, quality of care, organization, and the image of the institution.

Failure Mode and Effects Analysis for Risk Measurement

The FMEA (failure mode and effects analysis) approach was used to measure risks. For each EBIOS threat, we built scenarios that could be applicable in the hospital. For each scenario, we identified the potential risk (Fig. 3) and the impact (Fig. 4) for those involved. To do so, we used the following scales:

This allows the calculation of a risk criticality index: risk level = P * I. If the risk level is greater than or equal to 6, this means that the level of risk is not acceptable and it is necessary to implement corrective and/or preventive actions in order to obtain an acceptable residual risk level. Each action is also linked to one or more ISO 27002 recommendations. On completion of our risk assessment and taking into consideration all EBIOS threats, the recommended actions enable us to comply with the recommendations of ISO 27002. Our approach subsequently enabled us to formalize and deploy the information system security policy, upon validation by the management. It also enabled us to create the information system ACP (continuity plan) and ARP (plan for resuming activities) using tools provided by the consulting firm.

Low (unlikely)	Medium (possible)	High (likely)	Very high (frequent)
1 (< 3%)	2 (≤ 3% and < 20%)	3 (≤ 20% and < 30%)	4 (≥ 30%)

Fig. 3 Probability (P) of a scenario occurring

Classification	Level	Quality of care	Financial consequences	Institution's image	Legal consequences	Organization
Minor consequences	1	Minor prejudice, inconvenience	Low financial losses	Insignificant	Insignificant	Minor inconvenience for users
Major consequences	2	No bodily injury, moral prejudice	Loss ≤ 1% of turnover	Image suffers with regard to public	Risk of minor infringements of regulations	Significant inconvenience for users
Critical consequences	3	Serious bodily injury for patients	Loss ≤ 10% of turnover	Image suffers with regard to stakeholders	Risk of major infringements of regulations	Serious malfunction for departments and decision makers
Catastrophic consequences	4	Patient deaths	Loss > 10% of turnover	Criminal or civil proceedings	Risk of penalties	Discontinuation of critical activities

Fig. 4 Impact (I) of a scenario occurring

Conclusion

The approach implemented provided preliminary results that need to be consolidated through actions and new assessments of risk levels. The success and sustainability of this approach is based on a high level of determination and commitment from the management. The French national "digital hospital" program is also a very good lever for deployment. Its purpose is to:

- Coordinate all stakeholders (healthcare facilities, regional healthcare agencies (ARS), industrial companies, etc.) around a common roadmap for the management of hospital information systems;
- Ensure that the information systems of all health facilities reach a sufficient level of maturity so as to significantly improve the quality and safety of care, and performance levels in key functional areas relating to data security;
- Support innovative projects.

To increase the level of maturity of our information system and to achieve a common foundation in order to improve the quality and safety of care, four levers have been defined:

- Governance: tangible interpretation of information system objectives from national to local level, through the involvement of all stakeholders;
- Skills: enhancement of staff and health professionals' competencies in various aspects of the information system;
- Offer: stimulation and structuring the supply of certified solutions supporting innovative projects;
- Financing: financing plans to achieve objectives.

The common roadmap is based on three prerequisite criteria (identity and movements, reliability and availability, and confidentiality) and five functional areas: "Imaging results, biology and pathology", "Electronic patient records and interoperability", "Electronic prescription to support care plans", "Scheduling of resources and patient calendars" and "Medical and economic control".

References

EBIOS, ANSSI (Agence National de la Sécurité des Système d'information). http://www.ssi.gouv.fr/entreprise/bonnes-pratiques/ (2015)
ISO 27002, AFNOR, http://www.afnor.org/ (2015), March 1st 2015
ISO 27005, AFNOR. http://www.afnor.org (2015)

Erratum to:

A Mean-Field Analysis for the Two-Tiered Healthcare Network Through Nonlinear Markov Processes

Na Li, Quan-Lin Li, and Rui-Na Fan

A. Matta et al. (eds.), *Health Care Systems Engineering for Scientists and Practitioners*, Springer Proceedings in Mathematics & Statistics 169,
DOI 10.1007/978-3-319-35132-2

DOI 10.1007/978-3-319-35132-2_20

The original version of this book was inadvertently published with an incorrect order of Authors in Chapter 6. The correct order is as given below

Na Li, Quan-Lin Li, and Rui-Na Fan

The updated online version of the original chapter can be found at http://dx.doi.org/10.1007/978-3-319-35132-2_6

A. Matta et al. (eds.), *Health Care Systems Engineering for Scientists and Practitioners*, Springer Proceedings in Mathematics & Statistics 169,
DOI 10.1007/978-3-319-35132-2_20

MIX
Papier aus verantwortungsvollen Quellen
Paper from responsible sources
FSC® C105338

If you have any concerns about our products,
you can contact us on
ProductSafety@springernature.com

In case Publisher is established outside the EU,
the EU authorized representative is:
Springer Nature Customer Service Center GmbH
Europaplatz 3, 69115 Heidelberg, Germany

Printed by Libri Plureos GmbH
in Hamburg, Germany